普通高等院校城乡规划专业"十三五"精品教材

城乡规划英语教程
English Course for Urban-Rural Planning

主 编 路 旭

华中科技大学出版社
中国·武汉

内 容 提 要

本书依据国内普通高等院校专业英语课程要求,共分为15个单元。本书注重理论与实践相结合,书中既有对田园城市、广亩城市等经典理论的原文选介,也有对香港2030规划远景与策略、纽约区划法、墨尔本知识城市等前沿应用案例的介绍与分析,通过这些经典英文文献阐释城市发展战略研究、生态城市设计、详细规划、城市交通等城乡规划学科的热点领域。本书旨在培养学生对专业英语的应用能力,增加学生对国内外城市规划、城市生态学方面的基本理论和发展动态的了解,拓宽学生的专业视野。

本书是一本专业性较强的教材类书籍,可作为城乡规划学专业的本科教材,也可为高等院校风景园林、景观设计、建筑学等专业的本科教学而服务,同时还可供城乡规划及相关领域的专业技术人员、城市研究爱好者参考和借鉴。

图书在版编目(CIP)数据

城乡规划英语教程/路旭主编. —武汉:华中科技大学出版社,2019.8
普通高等院校城乡规划专业"十三五"精品教材
ISBN 978-7-5680-5318-1

Ⅰ.①城… Ⅱ.①路… Ⅲ.①城乡规划-英语-高等学校-教材 Ⅳ.①TU98

中国版本图书馆 CIP 数据核字(2019)第 163409 号

城乡规划英语教程 Chengxiang Guihua Yingyu Jiaocheng	路 旭 主编

策划编辑:简晓思
责任编辑:梁 任
封面设计:王亚平
责任校对:张会军
责任监印:朱 玢

出版发行:	华中科技大学出版社(中国·武汉)	电话:(027)81321913	
	武汉市东湖新技术开发区华工科技园	邮编:430223	
排 版:	华中科技大学惠友文印中心		
印 刷:	武汉科源印刷设计有限公司		
开 本:	850mm×1065mm 1/16		
印 张:	12.5		
字 数:	172千字		
版 次:	2019年8月第1版第1次印刷		
定 价:	39.80元		

本书若有印装质量问题,请向出版社营销中心调换
全国免费服务热线:400-6679-118 竭诚为您服务
版权所有 侵权必究

普通高等院校城乡规划专业"十三五"精品教材

总　　序

《管子》一书《权修》篇中有这样一段话："一年之计，莫如树谷；十年之计，莫如树木；百年之计，莫如树人。一树一获者，谷也；一树十获者，木也；一树百获者，人也。"这是管仲为富国强兵而重视培养人才的名言。

"十年树木，百年树人"即源于此。它的意思是说，培养人才是国家的百年大计，既十分重要，又不是短期内可以奏效的事。"百年树人"并不是非得一百年才能培养出人才，而是比喻培养人才的远大意义，要重视这方面的工作，并且要预先规划，长期、不间断地进行。

当前，我国城市和乡村发展形势迅猛，急缺大量的城乡规划专业应用型人才。全国各地设有城乡规划专业的学校众多，但能够既符合当前改革形势又适用于目前教学形式的优秀教材却很少。针对这种现状，急需推出一系列切合当前教育改革需要的高质量优秀专业教材，以推动应用型本科教育办学体制和运作机制的改革，提高教育的整体水平，并且有助于加快改进应用型本科办学模式、课程体系和教学方法，形成具有多元化特色的教育体系。

这套系列教材整体导向正确，科学精练，编排合理，指导性、学术性、实用性和可读性强。符合学校、学科的课程设置要求。以城乡规划学科专业指导委员会的专业培养目标为依据，注重教材的科学性、实用性、普适性，尽量满足同类专业院校的需求。教材内容上大力补充新知识、新技能、新工艺、新成果；注意理论教学与实践教学的搭配比例，结合目前教学课时减少的趋势适当调整了篇幅。根据教学大纲、学时、教学内容的要求，突出重点、难点，体现了建设"立体化"精品教材的宗旨。

这套系列教材以发展社会主义教育事业，振兴城乡规划类高等院校教育教学改革，促进城乡规划类高校教育教学质量的提高为己任，为发展我国高等城乡规划教育的理论、思想，对办学方针、体制，教育教学内容改革等进行了广泛深入的探讨，以提出新的理论、观点和主张。希望这套教材能够真实地体现我们的初衷，真正成为精品教材，受到大家的认可。

中国工程院院士　何镜堂

2007 年 5 月于北京

前　言

中国现代城乡规划起步于20世纪50年代中期,这60余年是中国城乡规划相关知识与西方相关知识体系不断融合的过程。在"一带一路"带动全球文明对话的今天,中国城乡规划正在向世界输送更多的理论、人才与实践经验,也在以前所未有的速度学习国际先进经验。掌握专业英语沟通能力与技巧已经成为当今城市规划师必备的专业技能之一。面对这样的情况,城乡规划专业英语课程在我国大专院校蓬勃开展,为提升广大学生的科研与实践能力奠定了坚实基础。

本人结合十余年的专业英语教学经验,不断进行总结和研究,思考本科生专业英语课程的目标定位与内容安排,在此基础上编写了本书。本书可帮助学生深入学习本专业经典理论与实践案例,使学生能够在英文语境下充分理解专业知识的初始内涵与生成背景,使其具备正确运用、拓展城乡规划专业知识的信心与能力。除此之外,本书旨在增加学生的专业词汇量,提高学生的精细阅读能力与表达能力,帮助学生建立完整的专业知识体系,以使学生能够从事国外专业文献的阅读与写作,并能够在工作和学术会议等情境下运用英文进行专业沟通。

本教材的主要特点如下。

①本书中所有文章均选自英文著作、论文、规划文本,力图让学生深入理解原作者的表述方式与思想内涵。

②本书尽量采用图文并茂的形式,对经典图示进行引用,便于广大学生直观理解。

③本书文章次序大致遵照城乡规划学科发展的时间线,同时也将内容难度逐渐提升,使学生经历一个由知识入门到实践应用的学习过程。

④本书突出沈阳建筑大学城乡规划学科特色,增加了城市生态规划方面的内容。

本书文章可以分为三种类型:第一种是理论溯源型,例如田园城市、广亩城市的节选或介绍,此类文章的语言蕴含了理论诞生之时的历史情境,可帮助学生深刻理解本学科经典理论的真实含义;第二类是案例介绍型,例如墨尔本知识城市、东京绿带等城市规划的经典案例,此类文章的语言较为通俗,涵盖了一些近年国际城市规划发展的新思潮与新动向;第三类是应用文类型,例如《香港2030规划远景与策略》《曼哈顿城市设计》《纽约区划手册》等规划文本或法规,此类文章在以往的城市规划英语教材中较为少见,本书增加的目的是让学生能够熟悉专业技术文体的表达方式,并掌握运用技巧。

本书共分为15个单元,其中第1~4、6、8、12~15单元由路旭编写,第5、7、9~11单元由付士磊编写,我校外语系教师王美瑜对全书进行了审读与修改,城乡规划专业研究生张晋熙、黄山、刘佳琦等协助开展了插图绘制和书稿校对等工作。华中科技大学出版社简晓思编辑、梁任编辑为本书的编审付出了辛勤的劳动。本书获得了华中科技大学出版社普通高等院校城乡规划专业"十三五"规划精品教材项目的支持,同时得到沈阳建筑大学2019年度一流本科教材立项支持。在此一并表示诚挚的谢意!

作者在编写本书过程中,虽然尽了很大努力,但由于水平和时间有限,难免存在疏漏和不当之处,请各位读者不吝赐教。

编　者

2019年7月

目　　录

Unit 1　Garden City ……………………………………………………（1）

Unit 2　Broadacre City: A New Community Plan ………………………（12）

Unit 3　The Uses of Sidewalk: Safety …………………………………（34）

Unit 4　The Four Functions of the City ………………………………（46）

Unit 5　Design With Nature ……………………………………………（58）

Unit 6　Accessibility ……………………………………………………（65）

Unit 7　Ecological Planning Methods …………………………………（77）

Unit 8　Patrick Geddes' Visual Thinking ………………………………（84）

Unit 9　Melbourne's Knowledge-based Urban Development …………（101）

Unit 10　Greenbelt of Tokyo ……………………………………………（113）

Unit 11　Planning Policy: The London Plan …………………………（121）

Unit 12　Midtown Manhattan's Office Projections ……………………（130）

Unit 13　The European Spatial Development Perspective ……………（145）

Unit 14　Reinforce Hub Functions of Hong Kong ……………………（159）

Unit 15　Residence Districts in NYC Zoning …………………………（169）

Reference Answer …………………………………………………………（184）

Unit 1   Garden City

This chapter was published in its original form as:

Ebenezer Howard (1902), Garden Cities of Tomorrow. London: Swan Sonnenschein & Co., 21-26.

TEXT

The objects of this land purchase may be stated in various ways, but it is sufficient here to say that some of the chief objects are these: to find for our industrial population work at wages of higher purchasing power, and to secure healthier surroundings and more regular employment. To manufacturers, co-operative societies, architects, engineers, builders, and mechanicians of all kinds, as well as to many engaged in various professions, it is intended to offer a means of securing new and better employment for their capital and talents, while to the agriculturists at present on the estate, as well as to those who may migrate thither, it is designed to open a new market for their produce close to their doors. Its object is, in short, to raise the standard of health and comfort of all true workers of whatever grade—the means by which these objects are to be achieved being a healthy, natural and economic combination of town and country life, and this on land owned by the municipality.

Garden City, which is to be built near the center of the 6,000 acres, covers an area of 1,000 acres, or a sixth part of the 6,000 acres, and might be of circular form, 1,240 yards (or nearly three-quarters of a mile) from center to circumference.

Six magnificent boulevards—each 120 feet wide—traverse the city from center to circumference, dividing it into six equal parts or wards. In the center is a circular space containing about five and a half acres, laid out as a beautiful and well-watered garden; and, surrounding this garden, each standing in its own ample grounds, are the larger public buildings—town hall, principal concert and lecture hall, theatre, library, museum, picture-gallery, and hospital.

The rest of the large space encircled by the "Crystal Palace" is a public park, containing 145 acres, which includes ample recreation grounds within very easy access of all the people.

Running all round the Central Park (except where it is intersected by the boulevards) is a wide glass arcade called the "Crystal Palace", opening on to the park. This building is in wet weather one of the favourite resorts of the people, whilst the knowledge that its bright shelter is ever close at hand tempts people into Central Park, even in the most doubtful of weathers. Here manufactured goods are exposed for sale, and here most of that class of shopping which requires the joy of deliberation and selection is done. The space enclosed by the Crystal Palace is, however, a good deal larger than is required for these purposes, and a considerable part of it is used as a Winter Garden—the whole forming a permanent exhibition of a most attractive character, whilst its circular form brings it near to every dweller in the town—the furthest removed inhabitant being within 600 yards.

Passing out of the Crystal Palace on our way to the outer ring of the town, we cross Fifth Avenue—lined, as are all the roads of the town, with trees—fronting which, and looking on to the Crystal Palace, we find a ring of very excellently-built houses, each standing in its own ample grounds; and, as we continue our walk, we observe that the houses are for the most part built either in con-centric rings, facing the various avenues (as the circular roads are termed), or fronting the boulevards and roads, which all converge to the center of the town. Asking the friend who accompanies us on our journey what the population of this little city may be, we are told about 30,000 in the city itself, and about 2,000 in the agricultural estate, and that there are in the town 5,500 building lots of an average size of 20 feet × 130 feet—the minimum space allotted for the purpose being 20 feet × 100 feet. Noticing the very varied architecture and design which the houses and groups of houses display—some having common gardens and co-operative kitchens—we learn that general observance of street line or harmonious departure from it are the chief points as to house-building over which the municipal authorities exercise control, for, though proper sanitary arrangements are strictly enforced, the fullest measure of individual taste and preference is encouraged.

Walking still toward the outskirts of the town, we come upon "Grand Avenue". This avenue is fully entitled to the name it bears, for it is 420 feet wide, and, forming a belt of green upwards of three miles long, divides that part of the town which lies outside Central Park into two belts. It really constitutes an additional park of 115 acres—a park which is within 240 yards of the furthest removed inhabitant. In this splendid avenue six sites, each of four acres, are occupied by public schools and their surrounding play-grounds and gardens, while other sites are reserved for churches, of such denominations as the religious beliefs of the people may determine, to be erected and maintained out of the funds of the worshippers and their friends. We observe that the houses fronting on Grand Avenue have departed from the general plan of concentric rings, and, in order to ensure a longer line of frontage on Grand Avenue, are arranged in crescents—thus also to the eye yet further enlarging the already splendid width of Grand Avenue.

On the outer ring of the town are factories, warehouses, dairies, markets, coal yards, timber yards, etc. All <u>fronting on</u> the circle railway, which encompasses the whole town, and which has sidings connecting it with a main line of railway which passes through the estate. This arrangement enables goods to be loaded direct into trucks from the warehouses and workshops, and so sent by railway to distant markets, or to be taken direct from the trucks into the warehouses or factories; thus not only effecting a very great saving in regard to packing and cartage, and reducing to a minimum loss from breakage, but also, by reducing the traffic on the roads of the town, lessening to a very marked extent the cost of their maintenance. The smoke fiend is kept well within bounds in Garden City; for all machinery is driven by electric energy, with the result that the cost of electricity for lighting and other purposes is greatly reduced.

The refuse of the town is utilized on the agricultural portions of the estate, which are held by various individuals in large farms, small holdings, allotments, cow pastures, etc. the natural competition of these various methods of agriculture, tested by the willingness of occupiers to offer the highest rent to the municipality, tending to bring about the best system of husbandry, or, what is more probable, the best systems adapted for various purposes. Thus it is easily conceivable that it may prove advantageous to grow wheat in very large fields, involving united action under a capitalist farmer, or by a body of co-operators; while the cultivation of vegetables, fruits, and flowers, which requires closer and more personal care, and more of the artistic and inventive faculty, may possibly be best dealt with by individuals, or by small groups of individuals having a common belief in the efficacy and value of certain dressings, methods of culture, or artificial and natural surroundings.

This plan, or, if the reader be pleased to so term it, this absence of plan, avoids the dangers of stagnation or dead level, and, though encouraging individual initiative, permits of the fullest co-operation, while the increased rents which follow from this form of competition are common or municipal property, and by far the larger part of them are expended in permanent improvements.

TRANSLATION

田园城市

这种购买土地的方式有许多目的，但其中最突出的目的是：促进工业人口形成更高的购买力，确保更健康的环境和更普遍的就业岗位。对于制造商、合伙人、建筑师、工程师、建筑商和机械师以及其他从事各种职业的人才资本来说，(这种方式)为他们提供了一种新的和更好的就业机会，同时对于目前在该用地上的农业学家，以及那些可能迁移到那里的农民，(这种方式)为靠近门的产品开辟了一个新的市场。总而言之,其意图在于提高所有各阶层忠实劳动者的健康和舒适水平——实现这些意图的手段就是把城市和乡村生活的健康、自然、经济因素组合在一起,并在这个城市的土地上体现出来。

田园城市建在6000英亩土地的中心附近,用地为1000英亩,占6000英亩土地的1/6。城市形状可以是圆形的,从中心到边缘为1240码(大约3/4英里)。

6条壮丽的林荫大道(boulevards)——每条宽120英尺——从中心通向四周,把城市划成6个相等的分区。中心是一块5.5英亩的圆形空间,布置成一个灌溉良好的美丽的花园;花园的四周环绕着用地宽敞的大型公共建筑——市政厅、音乐演讲大厅、剧院、图书馆、展览馆、画廊和医院。

其余的广大空间是一个用"水晶宫"(Crystal Palace)包围起来的"中央公园"(Central Park),面积为145英亩。它有宽敞的游憩用地,全体居民都能非常方便地享用。

环绕中央公园(不包括被林荫大道穿过的部分)是一个面向公园的宽敞的玻璃连拱廊,叫作"水晶宫",这是居民在雨天喜爱去的地方之一。晶莹透亮的建筑物中的信息近在咫尺,即使在最恶劣的天气也能吸引居民来到中央公园。工厂的产品在这里陈设出售,可供顾客尽情精心挑选。水晶宫的容量比购货活动所需的空间要大得多。它的绝大部分是作为"冬季花园"(Winter Garden)——整个水晶宫构成一个最有魅力的永久性展览会,它的环形布局使它能接近每一个城市居民——最远的居民也在600码以内。

出水晶宫向城外走,我们跨过五号大街(Fifth Avenue)——它和城内的所有道路一样,植有成行的树木——沿着这条大街,面向水晶宫,是一圈非常好的住宅,每所住宅都有宽敞的用地;如果继续前进,我们可以看到大多数住宅或者以同心圆方式面向各条大街(avenues,环路都称为大街),或者面向林荫大道和向城市中心汇聚的道路。陪同我们观光的朋友告诉我们,这座小城市自身有大约3万人,在农业用地上还有大约2000人。城内有5500块住宅建筑用地,其平均面积为20英尺×130英尺——最小面积为20英尺×100英尺。我们看到住宅和住宅组群(有些住宅共用花园和厨房)的装饰重点一般都沿着街道线或者适当退后于街道线,建筑设计手法千变万化,市政当局对此实行控制,既严格规定必要的卫生标准,又鼓励独具匠心充分反映个人的兴趣和爱好。

再向城外,我们来到"宏伟大街"(Grand Avenue)。这条大街真是名副其实,宽度为420英尺,形成一条长达3英里的带形绿地,把中央公园外围的城市地区划分成两条环带。它实际上构成了一个115英亩的公园——这个公园与最远的居民相距不到240码。在这条壮丽的大街上有6所公立学校,每所占地4英亩,设有游戏场和花园。大街的其他位置可供各种宗教信仰的居民建设各种派别的教堂,其建设费和维护费来自各派信徒和支持者筹集的基金。我们看到,面临宏伟大街的房屋并没有顺着总图确定的同心圆布置,而是按新月形布置,这样既增加了临街线的长度,又能使已经十分壮丽的宏伟大街在视觉上显得更宽阔。

在城市的外环有工厂、仓库、牛奶房、市场、煤场、木材场等等,它们都靠近围绕城市的环形铁路。环形铁路有侧线与通过该城市的铁路干线相连接。这种布局使得各种货物能够直接从仓库、车间装上货车,经由铁路运往远处的市场或者把货物直接从货车卸入仓库或工厂。这不仅大大节省了包装运输费用,尽可能减少破损,而且减少了城市道路上的交通量,从而明显地减少了道路的维护费。在田园城市中严格控制了烟尘的危害,因为所有机器都由电力驱动,这样又使照明和其他用电的费用大大降低。

城市的垃圾被用于当地的农业用地。这些农业用地分别属于大农场、小农户、自留地、牛奶场等单位。各类业主自愿地探索能向市政当局提供最高租金的农业经营方式。这些方式之间的自然竞争会带来最好的耕作体制，或者适应各种目的的较为可取的最好体制。不难设想，实践也许会证明粮食适于大面积种植，例如由一位农业资本家统管，或者由一个合作机构统管，而蔬菜、水果、花卉的种植，则要求较细致认真的管理，并具备较高的艺术修养和创造才能，可能最好由个人经营，或者由对某种经营方式、栽培方法或人为环境和自然环境的功效和价值有共同信念的个人组成的小团体来经营。

这种规划，如果读者愿意也可以称为留待规划的空白，可以避免经营上的停滞，而且，通过鼓励个人的创新，容许最完美的合作；而这种竞争方式所带来的增长的租金是属于公共的或市政的财产，其中绝大部分被用于长远的改进。

VOCABULARY

- wage [weɪdʒ] *n.* 工资；工钱；报酬
- enterprise [ˈentəpraɪz] *n.* 组织；公司；企业；计划；事业；事业心
- migrate [maɪˈɡreɪt] *vi.* 迁徙，移栖；(指人)大批外出
- thither [ˈðɪðə] *adv.* 到那儿，向那儿；在那儿
- municipality [mjuːnɪsɪˈpæləti] *n.* 市，自治市；市政府，市政当局
- circular [ˈsɜːkjələ] *adj.* 圆形的，环形的
- circumference [səˈkʌmfərəns] *n.* 圆周；周长；周线
- boulevard [ˈbuːləvɑːd] *n.* 大道，大街
- traverse [trəˈvɜːs] *vt.* 穿越；穿过
- ample [ˈæmpəl] *adj.* 足够的，充足的，充裕的
- encircle [ɪnˈsɜːkəl] *vt.* 围绕；环绕
- recreation [ˌrekriˈeɪʃən] *n.* 娱乐；消遣(方式)；重做；再现

- arcade [ɑːˈkeɪd] *n*. 拱廊,拱廊通道;拱廊商业街
- whilst [waɪlst] *conj*. 同时;时时,有时;当……的时候
- deliberation [dɪˌlɪbəˈreɪʃən] *n*. 考虑;讨论;从容;审慎
- concentric [kənˈsentrɪk] *adj*. （指圆）同心的,同轴的
- outskirt [ˈaʊtskɜːt] *n*. 郊区,市郊
- denomination [dɪˌnɒmɪˈneɪʃən] *n*. 分支,派别;面值,面额
- frontage [ˈfrʌntɪdʒ] *n*. （建筑物的）临街（或河）正面;临街（或河）地界
- crescent [ˈkresənt] *n*. 新月形(物),弦月形(物),蛾眉月形(物)
- warehouse [ˈweəhaʊs] *n*. 货仓,仓库;仓库批发店,大型零售商店
- cartage [ˈkɑːtɪdʒ] *n*. 货车运输;运费
- lessen [ˈlesən] *vt*. 减少,降低,减轻 *vi*. 减少,降低,减轻
- utilize [ˈjuːtəlaɪz] *vt*. 使用;利用;应用
- allotment [əˈlɒtmənt] *n*. 小块土地;分配,份额
- pasture [ˈpɑːstʃər] *n*. 牧场
- husbandry [ˈhʌzbəndrɪ] *n*. 种植;养殖;妥善使用,节俭使用
- faculty [ˈfækəltɪ] *n*. 能力,才能
- stagnation [stægˈneɪʃən] *n*. 停滞
- municipal [mjuːˈnɪsɪpəl] *adj*. 市政的;市立的;市政当局

EXERCISES

[Choose the right answers]

1. The centre of Garden City _____.

A. covers an area of 1,000 acres, or a sixth part of the 6,000 acres

B. is divided into six equal parts or wards

C. which contains the main function is green space

D. is the site of different kinds of larger public buildings, such as hospital

2. _____, the Crystal Palace can be it near to every dweller.

A. Because manufactured goods are exposed for sale

B. Because it is used as a Winter Garden

C. Because of its circular form

D. Because the furthest removed inhabitant is within 600 yards

3. According to the description of the street view of Fifth Avenue, we can know the information below except _____.

A. Fifth Avenue with trees-fronting

B. excellently-built houses arranged along Fifth Avenue

C. that the houses along Fifth Avenue are all built in con-centric rings

D. that through Fifth Avenue we can arrive at Crystal Palace

4. What's the difference between Grand Avenue's new layout and the general plan? _____.

A. Houses fronting on Grand Avenue are in concentric rings

B. The new layout aims to ensure a longer line of frontage on Grand Avenue

C. Houses are arranged in crescents

D. The splendid width of Grand Avenue is already large enough.

5. The underlined phrase "fronting on" refers to _____.

A. arrange along the road

B. face up

C. set forward

D. joint together

[Speak your mind]

1. What kind of measures do Garden City take to realize high standard of living conditions for citizens?

2. What's the orientation relationship between Crystal Palace and Central Park?

3. Is design method of residential groups in the town diverse or fixed?

4. Which avenue separates Central Park periphery into two belts?

5. Why did the cost of electricity significantly decrease?

Unit 2　Broadacre City: A New Community Plan

This chapter was published in its original form as:

Frank Lloyd Wright (1935), Broadacre City: A New Community Plan. Architectural Record. 344-349.

TEXT

Editors' introduction: for more than half a century, the question "who is the greatest American architect?" could have only one answer: Frank Lloyd Wright (1867—1959). First with his revolutionary "prairie houses" that seemed to grow directly out of the Midwest landscape with their long, low cantilevered rooflines, and later with such masterpieces as the Imperial Hotel in Tokyo, the Guggenheim Museum of Art in New York, and the breathtaking "Falling Water" in Western Pennsylvania, Wright became the spokesman for "organic architecture" and a style of building that expressed "the nature of the materials".

To many, Wright's architecture and "the architecture of American democracy" were synonymous. As an unabashed egotist and a pioneer in the field of media celebrity. Wright encouraged the popular identification of himself with the American spirit. He cultivated an imperious image of plain-speaking anti-collectivist democracy and sought personally to embody the notion of radical individualism. As an artistic genius, Wright despised the popular philistinism of his day and attributed the observable decline of American popular culture to "the mobocracy" and to the unprincipled bankers and politicians who served its interests. By the 1920s and 1930s, Wright had become a social revolutionary but not characteristically, of the socialist Left. Rather, Wright called for a radical transformation of American society to restore earlier Emersonian and Jeffersonian virtues. The physical embodiment of that utopia vision was Broadacre City.

Wright unveiled his model of Broadacre City, illustrated in Plate 29 at Rockefeller Centre, New York, in 1935. The article reprinted here represents his first and clearest statement of the revolutionary proposal whereby every citizen of the United States would be given a minimum of one acre of land per person, with the family homestead being the basis of civilization, and with government reduced to nothing more than a county architect who would be in charge of directing land allotments and the construction of basic community facilities. Many at the time thought the idea was totally outlandish, but Broadacre (and the small, efficient "Usonian" house) proved to be prophetic as sprawling suburban regions transformed the American landscape during the second half of the twentieth century.

Wright believed that two inventions—the telephone and the automobile—made the old cities "no longer modern" and he fervently looked forward to the day when dense, crowded conglomerations like New York and Chicago would wither and decay. In their place, Americans would reinhabit the rural landscape (and re-acquire the rural virtues of individual freedom and self-reliance) with a "city" of independent homesteads in which people would be isolated enough from one another to insure family stability but connected enough, through modern telecommunications and transportation, to achieve a real sense of community. Borrowing an idea from the anarchist philosopher Kropotkin, Wright believed that the citizens of Broadacre would pursue a combination of manual and intellectual work every day thus achieving a human wholeness that modern society and the modern city had destroyed. He also believed that a system of personal freedom and dignity through land ownership was the way to guarantee social harmony and avoid class struggle.

Broadacre City invites immediate comparison with the very different models of Ebenezer Howards Garden City and Le Corbusier's cities based on towers in a park. Intriguingly, the overall population density of Broadacre, on the one hand, and the garden city and Corbusian visions on the other, were not all that different, depending on the actual acreage of the surrounding parkland or greenbelt. And both Wright's and Le Corbusier's plans are wedded to the automobile one vision seeing a centralizing, the other a decentralizing effect. But the most revealing comparisons are with Robert Fishman's description of the now-emerging "technoburbs" and Melvin Webber's prediction of "a post-urban age". One cannot help but wonder whether what seemed impossible in 1935 may actually be realized, with the help of computer-based telecommunications and the possibility of "telecommuting" to work over the Internet, in the twenty-first century.

Given the simple exercise of several inherently just rights of man, the freedom to decentralize, to redistribute and to correlate the properties of the life of man on earth to his birthright—the ground itself—and Broadacre City becomes reality. As I see Architecture, the best architect is he who will devise forms nearest organic as features of human growth by way of changes natural to that growth. Civilization is itself inevitably a form but not, if democracy is sanity, is it necessarily the fixation called "academic". All regimentation is a form of death which may sometimes serve life but more often imposes upon it. In Broadacres all is symmetrical but it is seldom obviously and never academically so.

Whatever forms issue are capable of normal growth without destruction of such pattern as they may have. Nor is there much obvious repetition in the new city. Where regiment and row serve the general harmony of arrangement both are present, but generally, both are absent except where planting and cultivation are naturally a process or walls afford a desired seclusion. Rhythm is the substitute for such repetitions everywhere. Wherever repetition (standardization) enters, it has been modified by inner rhythms either by art or by nature as it must, to be of any lasting human value.

The three major inventions already at work building Broadacres, whether the powers that over-built the old cities otherwise like it or not.

1. The motor car: general mobilization of the human being.

2. Radio telephone and telegraph: electrical inter-communication becoming complete.

3. Standardized machine-shop production: machine invention plus scientific discovery.

The price of the major three to America has been the exploitation we see everywhere around us in waste and in ugly scaffolding that may now be thrown away. The price has not been so great if by way of popular government we are able to exercise the use of three inherent rights of any man.

1. His social right to a direct medium of exchange in place of gold as a commodity: some form of social credit.

2. His social right to his place on the ground as he has had it in the sun and air: land to be held only by use and improvements.

3. His social right to the ideas by which and for which he lives: public ownership of invention and scientific discoveries that concern the life of the people.

The only assumption made by Broadacres as ideal is that these three rights will be the citizen's so soon as the folly of endeavoring to cheat him of their democratic values becomes apparent to those who hold (feudal survivors or survivals), as it is becoming apparent to the thinking people who are held blindly abject or subject against their will.

The landlord is no happier than the tenant. The speculator can no longer win much at a game about played out. The present success-ideal, placing, as it does, premiums upon the wolf, the fox and the rat in human affairs and above all, upon the parasite, is growing more evident every day as a falsity just as injurious to the "successful" as to the victims of such success. Well-sociologically, Broadacres is release from all that fatal "success" which is, after all, only excess. So I have called it a new freedom for living in America. It has thrown the scaffolding aside. It sets up a new ideal of success.

In Broadacres, by elimination of cities and towns the present curse of petty and minor officialdom, government, has been reduced to one minor government for each county. The waste motion, the back and forth haul, that today makes so much idle business is gone. Distribution becomes automatic and direct, taking place mostly in the region of origin. Methods of distribution of everything are simple and direct. From the maker to the consumer by the most direct route.

Coal (one-third the tonnage of the haul of our railways) is eliminated by burning it at the mines and transferring that power, making it easier to take over the great railroad rights of way; to take off the cumbersome rolling stock and put the right of way into general service as the great arterial on which truck traffic is concentrated on lower side lanes, many lanes of speed traffic above and monorail speed trains at the center, continuously running. Because traffic may take off or take on at any given point, these arterials are traffic not dated but fluescent, and the great arterial as well as all the highways become great architecture, automatically affording within their structure all necessary storage facilities of raw materials, the elimination of all unsightly piles of raw material.

In the hands of the state but by way of the county, is all redistribution of land—a minimum of one acre going to the childless family and more to the larger family as effected by the state. The agent of the state in all matters of land allotment or improvement, or in matters affecting the harmony of the whole, is the architect. All building is subject to his sense of the whole as organic architecture. Here architecture landscape and landscape takes on the character of architecture by way of the simple process of cultivation.

All public utilities are concentrated in the hands of the state and county government as are matters of administration, patrol, fire post, banking, license and record, making politics a vital matter to everyone in the new city instead of the old case where hopeless indifference makes "politics" a grafter's profession.

In the buildings for Broadacres no distinction exists between much and little more and less. Quality is in all, for all, alike. The thought entering into the first or last estate is of the best. What differs is only individuality and extent. There is nothing poor or mean in Broadacres.

Nor does Broadacres issue any dictum or see any finality in the matter either of pattern or style.

Organic character is style. Such style has myriad forms inherently good. Growth is possible to Broadacres as a fundamental form, not as mere accident of change but as integral pattern unfolding from within.

Here now may be seen the elemental units of our social structure: the correlated farm, the factory—its smoke and gases eliminated by burning coal at places of origin, the decentralized school, the various conditions of residence, the home offices, safe traffic, simplified government. All common interests take place in a simple coordination wherein all are employed little farms, little homes for industry, little factories, little schools, a little university going to the people mostly by way of their interest in the ground, little laboratories on their own ground for professional men, and the farm itself, notwithstanding its animals, becomes the most attractive unit of the city. The husbandry of animals at last is in decent association with them and with all else as well. True farm relief.

To build Broadacres as conceived would automatically end unemployment and all its evils forever. There would never be labor enough nor could under-consumption ever ensue. Whatever a man did would be done—obviously and directly—mostly by himself in his own interest under the most valuable inspiration and direction: under training, certainly, if necessary. Economic independence would be near, a subsistence certain; life varied and interesting.

Every kind of builder would be likely to have a jealous eye to the harmony of the whole within broad limits fixed by the county architect, an architect chosen by the county itself. Each county would thus naturally develop an individuality of its own. Architecture—in the broad sense— would thrive.

In an organic architecture the ground itself predetermines all features; the climate modifies them; available means limit them; function shapes them.

Form and function are one in Broadacres. But Broadacres is no finality! The model shows four square miles of a typical countryside developed on the acre as unit according to conditions in the temperate zone and accommodating some 1,400 families. It would swing north or swing south in type as conditions, climate and topography of the region changed.

In the model the emphasis has been placed upon diversity in unity, recognizing the necessity of cultivation as a need for formality in most of the planting. By a simple government subsidy certain specific acres or groups of acre units are, in every generation, planted to useful trees, meantime beautiful, giving privacy and various rural divisions. There are no rows of trees alongside the roads to shut out the view. Rows where they occur are perpendicular to the road or the trees are planted in groups. Useful trees like white pine, walnut, birch, beech, fir, would come to maturity as well as fruit and nut trees and they would come as a profitable crop meantime giving character, privacy and comfort to the whole city. The general park is a flowered meadow beside the stream and is bordered with ranks of trees, tiers gradually rising in height above the flowers at the ground level. A music-garden is sequestered from noise at one end. Much is made of general sports and festivals by way of the stadium, zoo, aquarium, arboretum and the arts.

The traffic problem has been given special attention, as the more mobilization is made a comfort and a facility the sooner will Broadacres arrive. Every Broadacre citizen has his own car. Multiple-lane highways make travel safe and enjoyable. There are no grade crossings nor left turns on grade. The road system and construction is such that no signals nor any lamp-posts need be seen. No ditches are alongside the roads. No curbs either. An inlaid purfling over which the car cannot come without damage to itself takes its place to protect the pedestrian.

In the affair of air transport, Broadacres rejects the present airplane and substitutes the self-contained mechanical unit that is sure to come: an aerator capable of rising straight up and by reversible rotors able to travel in any given direction under radio control at a maximum speed of, say, 200 miles an hour, and able to descend safely into the hexacomb from which it arose or anywhere else. By a doorstep if desired.

The only fixed transport trains kept on the arterial are the long-distance monorail cars traveling at a speed (already established in Germany) of 220 miles per hour. All other traffic is by motor car on the twelve lane levels or the triple truck lanes on the lower levels which have on both sides the advantage of delivery direct to warehousing or from warehouses to consumer. Local trucks may get to warehouse-storage on lower levels under the main arterial itself. A local truck road parallels the swifter lanes.

Houses in the new city are varied: make much of fireproof synthetic materials, factory-fabricated units adapted to free assembly and varied arrangement but do not neglect the older nature-materials wherever they are desired and available.

Glass is extensively used as are roofless rooms. The roof is used often as a trellis or a garden. But where glass is extensively used it is usually for domestic purposes in the shadow of protecting overhangs.

Copper for roofs is indicated generally on the model as a permanent cover capable of being worked in many appropriate ways and giving a general harmonious color effect to the whole.

Electricity, oil and gas are the only popular fuels. Each land allotment has a pit near the public lighting fixture where access to the three and to water and sewer may be had without tearing up the pavements.

The school problem is solved by segregating a group of low buildings in the interior spaces of the city where the children can go without crossing traffic. The school building group includes galleries for loan collections from the museum, a concert and lecture hall, small gardens for the children in small groups and well-lighted cubicles for individual outdoor study: there is a small zoo, large pools and green playgrounds.

This group is at the very center of the model and contains at its center the higher school adapted to the segregation of the students into small groups.

This tract of four miles square, by way of such liberal general allotment determined by acreage and type of ground, including apartment buildings and hotel facilities, provides for about 1,400 families at, say, an average of five or more persons to the family.

To reiterate: the basis of the whole is general decentralization as an applied principle and architectural reintegration of all units into one fabric; free use of the ground held only by use and improvements; public utilities and government itself owned by the people of Broadacre City; privacy on one's own ground for all and a fair means of subsistence for all by way of their own work on their own ground or in their own laboratory or in common offices serving the life of the whole.

There are too many details involved in the model of Broadacres to permit complete explanation. Study of the model itself is necessary study. Most details are explained by way of collateral models of the various types of construction shown: highway construction, left turns, crossovers, underpasses and various houses and public buildings.

Anyone studying the model should bear in mind the thesis upon which the design has been built by the Taliesin Fellowship, built carefully not as a finality in any sense but as an interpretation of the changes inevitable to our growth as a people and a nation.

Individuality established on such terms must thrive. Unwholesome life would get no encouragement and the ghastly heritage left by overcrowding in overdone ultra-capitalistic centers would be likely to disappear in three or four generations. The old success ideals having no chance at all, new ones more natural to the best in man would be given a fresh opportunity to develop naturally.

TRANSLATION

广亩城市:一种新型规划思想

作者介绍:半个多世纪以来,"谁是美国最伟大的建筑师?"这个问题的答案只有一个:弗兰克·劳埃德·赖特(1867—1959)。首先是他的革命性的"草原式住宅",草原式住宅似乎是直接从中西部的地形中生长出来的,悬挂着长长的、低矮的屋顶。之后赖特还完成了像东京帝国饭店,纽约的古根海姆美术馆,以及宾夕法尼亚州西部的令人叹为观止的"流水别墅"等杰出作品。赖特成为了强调"有机的秩序"和"材料的本性"的重要性的建筑风格的代言人。

对许多人来说,赖特的建筑和"美国民主时期建筑"风格相似。作为一个毫不掩饰的自我主义者和媒体名人领域的先驱。赖特用美国精神来强化自己面对大众的身份特征。他展现了一种自然的、反集中主义的民主建筑风格,并独立思考如何体现个人主义的概念。作为一个天才艺术家,赖特不随波逐流追求当时流行的庸俗主义,并将美国大众文化的明显衰落归咎于"暴政"以及毫无原则地谋取利益的银行家和政治家。20 世纪 20 年代及 30 年代,赖特已经成为非典型的社会党左派的社会革命者。之后赖特呼吁彻底改变美国社会,以恢复早期埃默森和杰斐逊时期的社会景象。这一乌托邦理念的物质体现就是广亩城市。

赖特于 1935 年在纽约洛克菲勒中心的第 29 号板上展示了他的广亩城市的模型。这里转载这篇文章,表达了他最先提出同时也是最明确的主张,即家园是文明的基础,每一个美国公民至少可以得到一英亩土地,而政府只不过是一名负责指导土地分配和基本社区设施建设的建筑师。当时许多人认为这个想法完全是异想天开,但事实证明,布罗德阿克(和小型并高效的"乌萨尼亚"住宅)是具有预见性的,因为在二十世纪后半叶,广阔的郊区地区改变了美国的景观。

赖特相信随着电话和汽车这两项发明的产生，旧城已经"不再现代化"，他坚信像纽约和芝加哥这样密集、拥挤的集中城市终会有衰败的那一天。到那时，美国人将通过建立一个拥有独立宅基地的"城市"来重建乡村景观（并重新获得精神自由和独立自主的乡村生活），在那里，家庭之间将保持一定距离以确保家庭的稳定，同时通过现代通信和交通，使人们有足够的联系，从而获得真正的社区感。赖特借鉴了无政府主义哲学家克罗波特金（Kropotkin）的观点，认为广亩城市的居民要保持体力劳动和脑力劳动的平衡，从而恢复已经被现代社会和现代城市所毁灭的人类的整体性。他还认为，通过土地所有权建立个人自由和尊严的制度，是保障社会和谐、避免阶级斗争的一种方式。

广亩城市明显不同于霍华德的田园城市以及柯布西耶以公园里的塔楼为基础的城市的模型。有趣的是，广亩城市与田园城市和柯布西耶主张的城市并不是完全不同的，它们的总体人口密度都取决于周围公园和绿地的实际面积。赖特和勒柯布西耶的计划都与汽车有关，其中一个愿景是集中化，另一个是分散化的效果。但（对广亩城市来说）相比更具启发的是罗伯特·菲什曼对新兴的"技术郊区"的描述和梅尔文·韦伯对后城市时代的预测。人们不禁想知道，在21世纪通过互联网工作实现的电信和远程办公的帮助下，那些1935年不可能实现的事情是否会真正实现。

人类有许多与生俱来的权利，例如分散生活的权利、重新选择生活地的权利，鉴于（广亩城市）实现了这些权利，同时将地球上的人的生命财产与土地使用权利联系起来，广亩城市可能成为现实。我眼中的建筑，最好的建筑师会将建筑形式设计成类似人类生长的有机形式，将自然状态转化为生长状态。文明本身必然是一种形式，但是如果民主是理智的，就没有必要执着于某种"理论"。所有的管制都是一种死去的形式，它有时会服务生命，但更多的是强加于它。在广亩城市，一切都是公平的，（管制在广亩城市中)并不明显,甚至在理论上从未被提及。

任何形式的主题都能够保证城市正常的生长,只要不破坏它们所应有的布局形式。新城的那种更为显著的重复性布局也不应该被破坏。组团形式和行列形式应当并存,以实现城市布局的总体和谐,但是当遇到种植栽培的自然过程或者被墙体有意阻隔的时候,两种形式就会被打破。韵律感可以在任何地方替代这种重复手法。无论在哪里,当重复手法(标准化)出现的时候,它都必然会被艺术的或者自然的内在韵律所限定,从而具有持久的人性价值。

无论那些曾经对旧城过度建设的掌权者们喜欢与否,三个主要发明都已经在支持广亩城市的发展。

1. 汽车:人类的基本交通工具。
2. 无线电话和电报:日益完善的电力通信。
3. 标准化机械车间生产:机械发明与科学发现。

美国为三种主要发明付出的代价便是我们周围随处可见的开采浪费,而这些丑陋的废料现在也许会被丢弃。如果我们能够通过民选政府来行使人的三项固有权利,代价就不会这么大。

1. 作为中介直接交换物品而无需通过黄金的社会权利:某种形式的社会信用。
2. 在他的土地上拥有阳光空气的社会权利:只有通过使用和改良才能占有土地。
3. 拥有他赖以生存和为之奋斗的思想的社会权利:将与人民生活相关的发明和科学发现实行公有化。

广亩城市的唯一理想假设是这三种权利将会是公民的权利,在那时以民主价值欺骗他人的愚蠢行为变得昭然若揭,同时对于那些被盲目违背自己意志的思想者也显而易见。

房东并不比房客快乐。在一场比赛中,投机者不可能一直获胜。打个比方说如同在自然社会中,对狼、狐狸和老鼠尤其是对寄生虫的不完全真实的价值定义,目前成功的模式,正在对这种虚伪的"价值"以及这种"价值"评价的受害者造成着伤害,而且越来越明显。从社会学的角度来看,广亩城市弱化了所有的重大"成功",毕竟,那只是虚假的评价,所以我(将广亩城市模式)称为自由生活的美国。它把框架抛到一边并树立了一种新的成功模型。

在广亩城市,为了去除城镇现有的琐事化和小官吏化问题,行政管辖已精简为各县设一个小政府。如今,那些多余的工作,无必要的重复以及多余的闲置工作都被取消了。一切分配方法简单而直接,采用从制造商到消费者最直接的途径运输。

煤炭(占国家铁路运输总量的三分之一)因在矿井燃烧转换为其他能源而被消耗,煤炭运输需要占用极大的铁路通行权。把笨重的机车卸下来改造为主要的交通干线,把公路的右方作为主要的交通要道,卡车的交通主要集中在较低的侧线上,上面有许多车道,中间有不断行驶的单轨高速列车。由于交通可能在任何特定的地点发生,这些交通动脉不是静止的,而是运动的。交通动脉和所有高速公路都是伟大的建筑,运用结构设计隐藏了不美观的设施并提供必要的原材料储存设施。

土地属于国家,但通过州政府的方式重新分配——至少 1 英亩土地归无子女家庭所有,更多土地归国家组织的大家庭所有。建筑师是国家在土地分配、改善城市和协调整体的一切事务的代理人。所有的建筑都服从于有机建筑的整体意识。建筑的特征是通过建筑和景观的有序组织过程来呈现的。

所有的公共设施都集中在州政府手中,包括行政、巡逻、消防、银行、管理和登记等都是如此,这使得政策对新城的每一个人来说都是至关重要的,而不是像过去那样,绝望的冷漠使"公务员"成为贪污者的职业。

在广亩城市的建筑里,对所有人来说,(个人拥有物品的)多少没有区别,质量才是最重要的。对于居住者来说,每个人所选择的都是最好的,不同的只是个性和程度,在广亩城市没有贫穷和低下。

广亩城市并没有任何的定义,也没有任何在模式或风格固定不变的情况下的最终结果。

有机是一种风格。这种风格有与生俱来的好处。生长是广亩城市的基本形式,不仅是偶然的改变,而且是从内部自然演变的模式。

可以看到,我们社会结构的基本单元是相互关联的农场、工厂,它们需要消除燃烧煤炭产生的烟雾和气体而不是直接排放,同时需提供集中的学校、多种类型的居住环境、职住结合的办公空间、安全的交通运输设施和简化的政府管理部门。一切协调都简单地基于共同利益的基础之上,人们被雇佣于小型农场、小型工业、小型工厂、小型学校、小型大学、小实验室,在他们自己的土地上为专业人士服务,这些都使大众获得土地利益变成了现实。而拥有多种动物的农场,成为了这个城市最吸引人的部分。

按照设想的那样建造广亩城市,将自然而然地永久性地终结失业和犯罪。劳动力不足以及消费不足的状况也永远不会发生。在宝贵的灵感指导下,一个人所做的任何事,都会直接和显著地实现自己的利益:有时在训练中完成的是必要的。经济独立,丰富和有趣的生活就在眼前。

每一个建造者都很可能对整个州域范围的限定产生不同看法,并会对州的建筑师(州选民自己所选择的建筑师)产生指导作用。因此,每个州都会自然而然地发展出自己的特色。从广义上讲,建筑将繁荣发展。

在有机建筑中,土地本身所拥有的特征会被气候所影响,被技术所限制,被其功能所塑造。

在广亩城市中,形式与功能是统一的。但是广亩城市并不是最终产物!广亩城市的模型根据温带地区的条件开发,展示了一个在以英亩为单位的四平方英里的乡村,容纳了大约1400户家庭。随着该地区的条件、气候和地形的变化,广亩城市将向北或向南扩张。

在该模型中,重点强调统一的多样性,认识到在种植过程中,种植形式多样性的重要性。在每一代人中,通过简单的政府补贴,在某些特定的英亩地区或一组英亩单元种植有用的树木,同时美化环境,提供隐私空间和各种农村部门所在地。路旁不会有成片的树木挡住视线。它们与道路垂直以行的形式出现或者成组种植。白松、胡桃树、桦树、山毛榉、冷杉以及成熟的水果树和坚果树,这些可用的树木,它们作为经济作物的同时会赋予城市特色性、隐私性和舒适性。一般的公园是小溪边有一片鲜花草地,紧挨着一排排树木,层层叠叠,高过地面上的花朵。音乐花园的一端是隔音的。体育活动和节日大多在体育场、动物园、水族馆、植物园和艺术馆进行。

交通问题已经得到了特别的关注,因为交通更加便捷与舒适,很快就能到达广亩城市。每个广亩城市的居民都有自己的车。多车道公路使旅行变得安全而愉快。道路没有平交道口,也没有左转弯。道路系统和建筑物上看不到标志牌和路灯杆。道路两旁没有壕沟也没有限制物。一种镶嵌在道路上的装饰品可以让汽车在不受损的情况下阻止其变道来保护行人。

在航空运输方面,广亩城市拒绝使用现有的飞机,而是用自主独立的机械来替代:能够直线上升的通风机和在无线电控制下能向任意方向行驶的可逆转子,其最大速度可达200英里每小时,而且可以安全到达任意位置或者其他任意地方,如果必要的话,直达家门口也不是完全不可能的事情。

在主干道上保留的唯一固定运输列车是长途单轨车,以每小时220英里的速度(在德国已经确定)行驶。其他的交通是行驶机动车辆实现的,机动车辆在十二车道水平或较低水平的三重卡车车道上行驶,道路两侧都拥有直接到仓库或者从仓库到消费者的优势。当地的卡车可以在主干道下方较低层实现仓储服务。当地的卡车路线与更快捷的车道平行。

新城市房屋种类繁多:工厂制造的单元适合自由组装和不同的布置,

应用了大量的防火合成材料,但在自然材料可使用的情况下,不要忽视旧的自然材料的利用。

玻璃被广泛应用于透明屋顶房屋。房顶常被作为特色花架或者花园。但是,在广泛使用玻璃的情况下,它通常用来与悬垂物结构混合布置。

铜屋顶通常作为一个适合各种情况下工作的永久性的盖子置于模型上,并给整体增添了一种和谐的色彩。

电力、石油和天然气是为数不多的受欢迎的燃料。每个照明设备用地附近都配有一个孔洞,这样可以在不破坏人行道的情况下,通向包括用水管和污水管在内的三个管道。

校园问题的解决方法是将城市的内部空间的一些低矮的建筑物隔离分开,让孩子们可以不用穿过交通线路上学。学校的建筑群包括从用以借阅收藏的博物馆、用以展览的画廊、一个音乐会和演讲用的大厅,供小群儿童使用的小花园,以及供个人户外学习用的光线充足的小隔间,还需有一个小型动物园,大水池和绿化操场。

这个建筑群处于模型的正中心,包括其中心的高等学校,将学生们隔离成一些小群体。

在这个四平方英里的地块内,这种由面积和土地类型(包括公寓大楼和旅馆设施)所决定的这种普遍的自由分配方式,可供1400个家庭使用,家庭平均人数应在5个人或以上。

重申:整体原则的基础有一条普遍适用原则——权利普及化,同时使所有单位的建筑重新融为一体。只有在使用和改善土地的情况下才可以免费使用土地;广亩城市的居民是政府及公共设施的拥有者,人人在自己的土地上享有隐私,人人都可以通过在自己的场所、实验室以及为社会服务的办公室工作而享有公平的生存权利。

关于广亩城市的模型,有太多的细节,这里不能做出完整的解释。对模型本身的研究是必要的。大多数细节都是通过各种建筑的附属模型来解释的：高速公路建设，左转弯，交叉路口，地下通道，各种房屋和公共建筑。

任何研究这个模型的人都应该记住这个由塔里耶森奖学金所提出的观点，虽然它并不是最终结果，但对于一个民族和国家，广亩城市是对我们必然的改变的描述。

在这些基础条件下的城市个性会繁荣发展。腐败的生活得不到任何鼓励，过度拥挤的极端资本主义中心所下遗留下来的严重问题很可能在三个或四个年代内消失。旧的理想方案根本没有机会得到发展，新的模型对于人们来说更自然，更容易得到新的发展机会。

VOCABULARY

- homestead ['həʊmsted] *n.* 农场，农庄，农田
- community [kə'mju:nətɪ] *n.* 社区；群体；社团，团体，界
- ownership ['əʊnəʃɪp] *n.* 所有权；物主身份；拥有
- landscape ['lændskeɪp] *n.* 乡间，野外；（尤指乡村的）风景，景色
- technoburbs ['teknəʊbɜ:bs] *n.* 技术郊区城镇；技术近郊区（指城市开发最快的地区）
- distribution [dɪstrɪ'bju:ʃən] *n.* 分发，散发；分配
- arterial [[ɑː'tɪərɪəl] *adj.* 干线的；像动脉的；动脉的
- organic [ɔː'gænɪk] *adj.* 有机的，不使用化肥的，生物的；有机物的
- utopian [juː'təʊpɪən] *adj.* 乌托邦似的；完美世界的；以乌托邦为目标的
- prophetic [prə'fetɪk] *adj.* 预言的，预示的
- facilities [fə'sɪlətɪz] *n.* 设施；设备；天资，才能；特色
- harmony ['hɑːmənɪ] *n.* 协调，和谐，一致；和声
- anarchist ['ænəkɪst] *n.* 无政府主义者

- conglomeration [kənˌɡlɒməˈreɪʃən] n. 聚集物；混合体
- density [ˈdensɪti] n. （人口等的）密度；（物质的）密度
- telecommuting [ˌtelɪkəˈmjuːtɪŋ] n. 远程办公
- decentralize [diːˈsentrəlaɪz] vt. 分散管理，分权管理；将（权力）下放
- symmetrical [sɪˈmetrɪkəl] adj. 匀称的，对称的
- academically [ˌækəˈdemɪkəli] adv. 学术上；学业上
- rhythm [ˈrɪðəm] n. 节奏，韵律，节律；（尤指自然界中的）规则变化，规律
- sociologically [ˌsəʊsiəˈlɒdʒɪkəli] adv. 在社会学上地；社会上
- scaffold [ˈskæfəʊld] n. 脚手架；吊架，吊篮；断头台；绞刑台
- officialdom [əˈfɪʃəldəm] n. （尤指政府中效率低下的）官僚
- structure [ˈstrʌktʃə] n. 结构；构造；机构；体系
- material [məˈtɪəriəl] n. 材料，原料；素材；资料；布，织物，料子
- cultivation [ˌkʌltɪˈveɪʃn] vt. 培养，培育；耕作；栽培；种植；建立，培养（关系）
- unfolding [ʌnˈfəʊld] n. 演变；伸展
- individuality [ˌɪndɪvɪdʒuˈæləti] n. 个性；特征，特质
- temperate [ˈtempərət] adj. （气候）温带的；（行为）温和的，有节制的
- diversity [daɪˈvɜːsəti] n. 多样性，多样化，差异，不同点
- ditch [dɪtʃ] n. 沟渠，壕沟
- lane [leɪn] n. 乡间小路；小巷，小街，胡同；车道
- storage [ˈstɔːrɪdʒ] n. 储存；储藏；储备
- fireproof [ˈfaɪəpruːf] adj. 耐火的；防火的

EXERCISES

[Choose the right answers]

1. Which of the following is not Wright's work? _____.

   A. Organic architecture

   B. The architecture of American democracy

   C. Guggenheim Museum of Art in New York

   D. "Falling Water" in Western Pennsylvania

2. Social right is one of the three inherent rights of any man, one who owns social right means _____.

   A. that direct exchange is allowed

   B. that he can exchange something by gold as a commodity

   C. that land is held by everyone like sun and air

   D. public ownership of any invention and scientific discoveries

3. By _____, Broadacres deal with the current problems facing cities and states.

   A. streamlining government departments

   B. eliminating of cities and towns

   C. subjoining minor officialdom

   D. improving government

4. According to elemental units of our social structure, _____.

   A. the correlated farm and the factory must eliminate smoke and gases they produced originally

   B. Broadacres might provide the decentralized school

   C. Broadacres provide a unified mode of residential area

   D. the home offices, safe traffic, simplified government is not necessary

5. According to the description of the traffic problem of Broadacres, we can know the information below except that _____.

A. private transportation is not necessary for every Broadacre citizen has his own car

B. Broadacres rejects the present airplane

C. the self-contained mechanical unit can achieve door-by-door transport

D. the only fixed transport trains are the long-distance monorail cars

[Speak your mind]

1. How much land would the United States citizens get per person at least as homestead?

2. What is the commonality of Broadacre City, Garden City and Corbusian visions?

3. What is the main emphasis in Broadacres?

4. What approach would be adopt to realize simple and direct distribution?

Unit 3 The Uses of Sidewalk: Safety

This chapter was published in its original form as:

Jacobs, J. (1961), "The Uses of Sidewalks: Safety", in Jacobs, J. (1961), The Death and Life of Great American Cities, Penguin, Harmondsworth, 39-65.

TEXT

Today barbarism has taken over many city streets, or people fear it has, which comes to much the same thing in the end. The barbarism and the real, not imagined, insecurity that gives rise to such fears cannot be tagged a problem of the slums. The problem is most serious, in fact, in genteel-looking "quiet residential areas" like that my friend was leaving.

It cannot be tagged as a problem of older parts of cities. The problem reaches its most baffling dimensions in some examples of rebuilt parts of cities, including supposedly the best examples of rebuilding, such as middle-income projects. The police precinct captain of a nationally admired project of this kind (admired by planners and lenders) has recently admonished residents not only about hang in garound outdoors after dark but has urged them never to answer their doors without knowing the caller. Life here has much in common with life for the three little pigs or the seven little kids of the nursery thrillers. The problem of sidewalk and doorstep insecurity is as serious in cities which have made conscientious efforts at rebuilding as it is in those cities that have lagged. Nor is it illuminating to tag minority groups, or the poor, or the outcast with responsibility for city danger. There are immense variations in the degree of civilization and safety found among such group sand among the city areas where they live. Some of the safest sidewalks in New York City, for example, at any time of day or night, are those along which poor people or minority groups live. And some of the most dangerous are in streets occupied by the same kinds of people. All this can also be said of other cities.

Deep and complicated social ills must lie behind delinquency and crime, in suburbs and towns as well as in great cities. This book will not go into speculation on the deeper reasons. It is sufficient, at this point, to say that if we are to maintain a city society that can diagnose and keep abreast of deeper social problems, the starting point must be, in any case, to strengthen whatever workable forces for maintaining safety and civilization do exist in the cities we do have. To build city districts that are custom made for easy crime is idiotic. Yet that is what we do.

The first thing to understand is that the public peace—the sidewalk and street peace—of cities is not kept primarily by the police, necessary as police are. It is kept primarily by an intricate, almost unconscious, network of voluntary controls and standards among the people themselves, and enforced by the people themselves. In some city areas—older public housing projects and streets with very high population turnover are often conspicuous examples—the keeping of public sidewalk law and order is left almost entirely to the police and special guards. Such places are jungles. No number of police can enforce civilization where the normal, casual enforcement of it has broken down.

The second thing to understand is that the problem of insecurity cannot be solved by spreading people out more thinly, trading the characteristics of cities for the characteristics of suburbs. If this could solve danger on the city streets, then Los Angeles should be a safe city, because superficially Los Angeles is almost all suburban.

It has virtually no districts compact enough to qualify as dense city areas. Yet Los Angeles cannot, any more than any other great city, evade the truth that, being a city, it is composed of strangers not all of whom are nice. Los Angeles's crime figures are flabbergasting. Among the seventeen standard metropolitan areas with populations over a million, Los Angeles stands so pre-eminent in crime that it is in a category by itself. And this is markedly true of crimes associated with personal attack, the crimes that make people fear the streets.

Here we come up against an all-important question about any city street: how much easy opportunity does it offer to crime? It may be that there is some absolute amount of crime in a given city, which will find an outlet somehow (I do not believe this). Whether this is so or not, different kinds of city streets garner radically different shares of barbarism and fear of barbarism.

Some city streets afford no opportunity to street barbarism. The streets of the North End of Boston are outstanding examples. They are probably as safe as any place on earth in this respect. Although most of the North End's residents are Italian or of Italian descent, the district's streets are also heavily and constantly used by people of every race and background. Some of the strangers from outside work in or close to the district; some come to shop and stroll; many, including members of minority groups who have inherited dangerous districts previously abandoned by others, make a point of cashing their pay-cheques in North End stores and immediately making their big weekly purchases in streets where they know they will not be parted from their money between the getting and the spending.

Meantime, in the Elm Hill Avenue section of Roxbury, a part of inner Boston that is suburban in superficial character, street assaults and the ever present possibility of more street assaults with no kibitzers to protect the victims, induce prudent people to stay off the sidewalks at night. Not surprisingly, for this and other reasons that are related (dispiritedness and dullness), most of Roxbury has run down. It has become a place to leave.

I do not wish to single out Roxbury or its once fine Elm Hill Avenue section especially as a vulnerable area; its disabilities, and especially its great blight of dullness, are all too common in other cities too. But differences like these in public safety within the same city are worth nothing. The Elm Hill Avenue section's basic troubles are not owing to a criminal or a discriminated against or a poverty-stricken population. Its troubles stem from the fact that it is physically quite unable to function safely and with related vitality as a city district. This is something everyone already knows: a well used city street is apt to be a safe street. A deserted city street is apt to be unsafe.

But how does this work, really? And what makes a city street well used or shunned? Why is the sidewalk mall in Washington Houses, which is supposed to be an attraction, shunned? Why are the sidewalks of the old city just to its west not shunned? What about streets that are busy part of the time and then empty abruptly? A city street equipped to handle strangers, and to make a safety asset, in itself, out of the presence of strangers, as the streets of successful city neighbour hoods always do, must have three main qualities.

First, there must be a clear demarcation between what is public space and what is private space. Public and private spaces cannot ooze into each other as they do typically in suburban settings or in projects.

Second, there must be eyes upon the street, eyes belonging to those we might call the natural proprietors of the street. The buildings on a street equipped to handle strangers and to ensure the safety of both residents and strangers must be oriented to the street. They cannot turn their backs or blank sides on it and leave it blind.

And third, the sidewalk must have users on it fairly continuously, both to add to the number of effective eyes on the street and to induce the people inbuildings along the street to watch the sidewalks insufficient numbers.

Nobody enjoys sitting on a stoop or looking out a window at an empty street. Almost nobody does such a thing. Large numbers of people entertain themselves, off and on, by watching street activity. In settlements that are smaller and simpler than big cities, controls on acceptable public behaviour, if not on crime, seem to operate with greater or lesser success through a web of reputation, gossip, approval, disapproval, and sanctions, all of which are powerful if people know each other and word travels.

But a city's streets, which must control not only the behaviour of the people of the city but also of visitors from suburbs and towns who want to have a big time away from the gossip and sanction sat home, have to operate by more direct, straightforward methods. It is a wonder cities have solved such an inherently difficult problem at all. And yet in many streets they do it magnificently.

It is futile to try to evade the issue of unsafe city streets by attempting to make some other features of a locality, say interior courtyards or sheltered play spaces, safe instead. By definition again, the streets of a city must do most of the job of handling strangers, for this is where strangers come and go. The streets must not only defend the city against predatory strangers, they must protect the many, many peaceable and well-meaning strangers who use them, ensuring their safety too as they pass through. Moreover, no normal person can spend his life in some artificial haven, and this includes children. Every one must use the streets.

TRANSLATION

人行道的使用：安全

今天野蛮行为已经侵袭了很多城市街道，或者说，人们对此类事件的担心最终导致了同样的后果。造成这种恐惧的野蛮行为，以及现实中的而非想象中的不安全现象，不能只归结于贫民窟的问题。事实上，这种问题发生在那种看上去很优雅的"安静的住宅区"是最严重的，就像我的朋友离开的那种地方。

也不能将之归咎于老城区的问题。这个问题在一些旧城重建的案例中达到了最难以解决的程度，包括那些在被认为是最好的重建项目，例如中产收入的项目，也是如此。一个负责此类（被规划师和投资商们誉为）全国先进项目的片区警长，告诫居民不要在天黑后出门闲逛，并且告诫居民不要在不认识对方的情况下给人开门。这里的生活竟然与三只小猪和七只小羊的恐怖故事相似。人行道和门阶的安全问题不仅在那些落后的城市中很严重，在那些大力重建的城市中同样严重。也不应该以城市安全责任为名，给少数群体、穷人或者流浪人士贴上标签。这些人居住的城市片区之间的文明程度及安全程度是大不相同的。例如，无论白天还是晚上任何时间，在纽约的一些安全的人行道两边是那些穷人和少数民族居住的地方。而那些十分危险的人行道也是被这些人占据。这种情况在其他城市也基本相同。

在郊区、城镇以及一些大城市中，在违法犯罪背后必然隐藏着深刻而复杂的社会弊病。这本书不会探究更深层次的原因。这一点足以说明：如果我们要维持一个能够诊断和及时了解更深层次社会问题的城市社会，出发点就应该放在加强所有可用的力量来维护城市中我们已经拥有的安全与文明上。建设那些为容易实施犯罪而定制的城区是十分愚蠢的行为。但我们现在就在这么做。

首先要弄明白的是城市的公共治安（人行道及街道的秩序）不是主要由警察来维持的，尽管警察是必要的。它主要是由一个复杂的、几乎无意识的网络来维持的，一个存在于人群之间的自愿性控制和行为标准，由人群自发执行。在一些城市地区（常见案例是旧的公共住房项目和人口流动率高的街道）维护公共人行道的法律和秩序几乎全部依赖于警察和特殊警卫。这样的地方像一个丛林一样危险。在一个常规的和非正式的执行力都被毁坏的地方，再多警察也不能提升其文明程度。

第二件要理解的事情是不安全的问题不能通过分散人口、降低密度，将城市特征替换为郊区特征来解决。如果这个可以解决城市街道的危险问题，洛杉矶已经是一个安全的城市了，因为从外表上看，洛杉矶已经几乎郊区化了。

事实上洛杉矶已经没有任何区域可以紧凑到被归入密集城市区域了。然而，与其他的城市一样，洛杉矶也不能避开这个事实，它是由很多的不全是好人的陌生人组成的一座城市。洛杉矶的犯罪数据让人瞠目结舌。在

17个人口超过百万的标准大都市地区中,洛杉矶的犯罪率是如此之高,以至于它被独自归为一类。明显的事实是,犯罪与人身攻击相关,犯罪使人惧怕街道。

这里,我们遇到一个有关于城市街道重要的问题:城市街道给犯罪提供了多少便利?也许在一座城市里就有一个固定数量的犯罪,它们将寻找某处来发生(我不相信这点)。无论是或不是,不同类型的城市街道从根本上滋生了不同比例的野蛮行为及对野蛮行为的恐惧。

一些城市并不给街道野蛮行为提供机会。波士顿北端的街道就是一个好案例。这里几乎能与地球上任何安全之处媲美。虽然大多数北端居民是意大利人或者意大利后裔,但其他各个种族和背景的人也经常和频繁地出现在这条街道上。一些外地的陌生人来到这个地区及附近工作,一些人来到这里购物及闲逛;很多人,包括一些居住在被别人抛弃的危险地带的少数群体来北方的商店兑换他们的薪水支票,并且立即在这购买他们一星期所需要的货物,因为他们知道在这个地方卖东西和买东西,他们的钱不会丢失。

同时,在罗克斯伯里的埃尔姆山大道区,一个表面上看是郊区的波士顿内城区域,经常有街头袭击事件发生却没有人来保护受害者,这使得一些小心谨慎的人在夜晚都远离街道。毫不奇怪,因为这样那样的相互关联的原因(缺少活力和单调乏味),大部分罗克斯伯里地区都破败了,成为了一个被背弃的地方。

我并不希望把罗克斯伯里或它曾经很好的埃尔姆山大道区作为一个易受袭击的区域列出来;它的缺陷,特别是它的萧条,和其他城市一样太常见了。但是在同一个城市,公共安全方面的此类差别却并不重要。埃尔姆山大道区域的根本问题不是因为存在着罪犯、歧视或者贫困人群。它的问题来源于一个事实:在实体空间上无法安全运转,它不像城市片区那样具有生命力。人所共知的是:一个较常使用的城市街道往往是安全的;一个荒废的城市街道往往是危险的。

但是这事实上是如何形成的?并且是什么让一个城市街道被充分利

用或者无人光顾呢？为什么华盛顿宫的人行道购物中心被认为是具有吸引力的，却无人光顾？还有那些有时候很热闹，有时候却突然空无一人的街道呢？一个有能力应付陌生人，成为本质上安全的资产、避开陌生人的街道，就像那些很成功的城市街区那样，必须要有三个主要的特点。

首先，公共空间和私人空间之间必须有明确的界限。公共空间和私人空间不能像在郊区或项目中那样相互渗透。

第二，必须有一双眼睛盯着街道，被我们称为街道的自然居住者的眼睛。街道两侧的建筑物应该朝向街道，有能力应对陌生人，并确保本地居民和陌生人的安全。这些建筑不能将背面或盲区朝向街道否则将失去保护的眼睛。

第三，街道上必须有适当的持续性的使用者，这样不仅可以增加街道上眼睛的数量，也可以增加人们在楼里看向街道的数量。

没人会喜欢坐在门廊里或者透过窗户看空荡荡的街道。几乎没有人会这样做。大部分人喜欢时不时地观看街头活动来娱乐自己。在一些比大城市更小更简单的聚落，在不构成犯罪的情况下，对可接受的公共行为的控制是通过一个由名誉、留言、赞许、反对和制止等行为构成的网络来运转的，如果人们彼此了解并且消息通畅，这将是很有用的方式。

但是城市街道需要控制的不仅是城市居民的行为，还有来自郊区和小城市，希望逃避家乡舆论和约束的外来者，因此必须要通过更明确、更直接的方法来控制。很难说城市是否解决了这个固有的难题。但是在很多街道，他们做得非常出色。

那些试图通过制造一些其他安全的地方性特征，如内部的庭院或者有遮蔽的游玩空间，来规避城市街道的不安全问题的做法是徒劳的。再次回到定义，城市街道必须做很多工作来应对陌生人，因为这里是陌生人往来的地方。街道不仅要防范侵略性的陌生人，还要保护很多使用街道的和平的、善意的陌生人，确保他们能安全通过。何况，没有人可以在与世隔绝的地方度日，包括孩子。每个人都需要使用街道。

VOCABULARY

- barbarism [ˈbɑːbərɪzəm] n. 野蛮行径；暴虐行径；粗俗行为
- slum [slʌm] n. （尤指城市中的）贫民窟，棚户区；脏乱的地方
- rebuilding [riˈbɪldɪŋ] n. 重建　v. 重建；使复原
- minority [maɪˈnɒrəti] n. 少数民族；少数派；少数民族人士；少数派人士；少数
- variation [ˌveəriˈeɪʃən] n. 变化；变动；变奏曲
- delinquency [dɪˈlɪŋkwənsi] n. （尤指年轻人的）不良行为，违法行为；没有按协定偿还债务
- suburb [ˈsʌbɜːb] n. 城郊；近郊住宅区
- civilization [ˌsɪvəlaɪˈzeɪʃən] n. （特定时期、特定社会或国家的）文明，文化；文明社会；教化；开化
- district [ˈdɪstrɪkt] n. 区，区域
- custom [ˈkʌstəm] n. 风俗，习俗；传统；习惯；惯例；（尤指对商店的）光顾，惠顾
- intricate [ˈɪntrɪkət] adj. 错综复杂的；复杂精细的；难理解的；难解决的
- unconscious [ʌnˈkɒnʃəs] adj. （想法或感情）无意识的，潜意识的；昏迷的，不省人事的
- suburban [səˈbɜːbən] adj. 近郊住宅区的；城郊的；平淡乏味的；无趣的
- compact [kəmˈpækt] adj. 紧凑的；密实的
- outlet [ˈaʊtlet] n. 出口，排放孔；（情绪或精力的）发泄方式，发泄途径，施展的机会
- descent [dɪˈsent] n. 血缘关系，家族关系
- previously [ˈpriːviəsli] adv. 以前地，先前地

- purchase [ˈpɜːtʃəs] *n.* 购买,采购
- assault [əˈsɒlt] *n.* 殴打;袭击,攻击;硬仗
- surprisingly [səˈpraɪzɪŋli] *adv.* 意外地,出人意料地
- vulnerable [ˈvʌlnərəbəl] *adj.* 易受伤的;易受影响(或攻击)的;脆弱的
- demarcation [diːmɑːˈkeɪʃən] *n.* (界线或规则)分界;分界线
- ooze [uːz] *vt.* 渗出;冒出;分泌出 *n.* (河或湖底的)淤泥,软泥,烂泥
- stoop [stuːp] *n.* 门廊,门阶;驼背,弓背
- gossip [ˈgɒsɪp] *n.* (有关别人隐私的)流言蜚语,闲言碎语,闲聊

EXERCISES

[Choose the right answers]

1. What is not mentioned in the description of barbarism in the article? _____.

    A. Barbarism has taken over many city streets

    B. People fear barbarism

    C. Barbarism promotes urban development

    D. Slums cannot be blamed

2. What is the wrong description about the order of urban public space? _____.

    A. It is kept primarily by an intricate

    B. It is kept primarily by the police

    C. Networks are created and controlled by people

    D. It is the duty of the police to maintain public order

3. Acoording to this article, how to solve the security problem? _____.

A. Spreading people out more thinly

B. Reducing population density

C. Trading the characteristics of cities for the characteristics of suburbs

D. Separating public and private spaces

4. In this article, which cities have safer streets? _____.

A. Los Angeles

B. The streets of the North End of Boston

C. Elm Hill Avenue section of Roxbury

D. Washington

5. Which of the following is not characteristic of a successful city block? _____.

A. Clear boundaries between public and private spaces

B. The width of the street should be more than ten meters

C. Street eye

D. The number of pedestrians increased

6. In this article, what is the wrong description of the role of city streets? _____.

A. Streets cannot monitor strangers

B. Control the behaviour of the people of the city

C. Defend the city against predatory strangers

D. Protect the many well-meaning strangers

[Speak your mind]

1. What do you make of the barbarism of city streets?

2. What do you think about sidewalk safety?

3. Have you ever read the life and death of a big American city? Tell me about your reading experience.

Unit 4   The Four Functions of the City

This chapter was published in its original form as:
Congress Internationaux d'Architecture moderne (CIAM), The Athens Charter, 1933.

TEXT

1. Dwelling

(1) The population density is too great in the historic, central districts of cities as well as in some nineteenth century areas of expansion: densities rise to 1000 and even 1500 inhabitants per hectare (approximately 400 to 600 per acre).

(2) In the congested urban areas, housing conditions are unhealthy due to insufficient space within the dwelling, absence of useable green spaces and neglected maintenance of the buildings (exploitation based on speculation). This situation is aggravated by the presence of a population with a very low standard of living, incapable of initiating ameliorations (mortality up to 20 per cent).

(3) Extensions of the city devour, bit by bit, its surrounding green areas. This ever greater separation from natural elements heightens the harmful effects of bad sanitary conditions.

(4) Dwellings are scattered throughout the city without consideration of sanitary requirements.

(5) The most densely populated districts are in the least favorable situations, on unfavorable slopes, invaded by fog or industrial emanations, subject to flooding, etc.

(6) Low in density developments occupy the advantageous sites, sheltered from unfavorable winds, with secure views opening onto an agreeable countryside, lake, sea, or mountains, etc. and with ample air and sunlight.

(7) This segregation of dwellings is sanctioned by custom, and by a system of local authority regulations considered quite justifiable: zoning.

(8) Buildings constructed alongside major routes and around crossroads are unsuitable for dwellings because of noise, dust and noxious gases.

(9) The traditional alignment of houses along the sides of roads means that good exposure to sunlight is only possible for a minimum number of dwellings.

(10) The distribution of community services related to housing is arbitrary.

(11) Schools, in particular, are frequently sited on busy traffic routes and too far from the houses they serve.

(12) Suburbs have developed without plans and without well organized links with the city.

(13) Attempts have been made too late to incorporate suburbs within the administrative unit of the city.

(14) Suburbs are often merely an agglomeration of hutments where it is difficult to collect funds for the necessary roads and services.

IT IS RECOMMENDED

(15) Residential areas should occupy the best places in the city from the point of view of typography, climate, sunlight and availability of green space.

(16) The selection of residential zones should be determined on grounds of health.

(17) Reasonable densities should be imposed related both to the type of housing and to the conditions of the site.

(18) A minimum number of hours of sunlight should be required for each dwelling unit.

(19) The alignment of housing along main traffic routes should be forbidden.

(20) Full use should be made of modern building techniques in constructing high rise apartments.

(21) Highrise apartments placed at wide distances apart liberate ground for large open spaces.

2. Recreation

(1) Open spaces are generally insufficient.

(2) When there is sufficient open space it is often badly distributed and therefore not readily usable by most of the population.

(3) Outlying open spaces cannot ameliorate areas of downtown congestion.

(4) The few sports fields, for reasons of accessibility, usually occupy sites earmarked for future development for housing or industry: which makes for a precarious existance and their frequent displacement.

(5) Land that could be used for week-end leisure is often very difficult of access.

IT IS RECOMMENDED

(6) All residential areas should be provided with sufficient open space to meet reasonable needs for recreation and active sports for children, adolescents and adults.

(7) Unsanitary slums should be demolished and replaced by open space. This would ameliorate the surrounding areas.

(8) The new open spaces should be used for well-defined purposes: children's playgrounds, schools, youth clubs and other community buildings closely related to housing.

(9) It should be possible to spend week-end free time in accessible and favorable places.

(10) These should be laid out as public parks, forests, sports grounds, stadiums, beaches, etc.

(11) Full advantages should be taken of existing natural features: rivers, forests, hills, mountains, valleys, lakes, sea, etc.

3. Work

(1) Places of work are no longer rationally distributed within the urban complex. This comprises industry, workshops, offices, government and commerce.

(2) Connections between dwelling and place of work are no longer reasonable: they impose excessively long journeys to work.

(3) The time spent in journeying to work has reached a critical situation.

(4) In the absence of planning programs, the uncontrolled growth of cities, lack of foresight, land speculation, etc. have caused industry to settle haphazardly, following no rule.

(5) Office buildings are concentrated in the downtown business

district which, as the most privileged part of the city, served by the most complete system of communications, readily falls prey to speculation. Since offices are private concerns effective planning for their best development is difficult.

(6) Distances between work places and dwelling places should be reduced to a minimum.

(7) Industrial sectors should be separated from residential sectors by an area of green open space.

(8) Industrial zones should be contiguous with railroads, canals and highways.

(9) Workshops, which are intimately related to urban life, and indeed derive from it, should occupy well designed areas in the interior of the city.

(10) Business districts devoted to administration both public and private, should be assured of good communications with residential areas as well as with industries and workshops within the city and upon its fringes.

4. Transportation

(1) The existing network of urban communications has arisen from an agglomeration of the aids roads of major traffic routes. In Europe these major routes date back well into the middle ages, sometimes even into antiquity.

(2) Devised for the use of pedestrians and horse drawn vehicles, they are inadequate for today's mechanized transportation.

(3) These inappropriate street dimensions prevent the effective use of mechanized vehicles at speeds corresponding to urban pressure.

(4) Distances between crossroads are too infrequent.

(5) Street widths are insufficient. Their widening is difficult and often ineffectual.

(6) Faced by the needs of high speed vehicles, present the apparently irrational street pattern lacks efficiency and flexibility, differentiation and order.

(7) Relics of a former pompous magnificence designed for special monumental effects often complicate traffic circulation.

(8) In many cases the railroad system presents a serious obstacle to well planned urban development. It barricades off certain residential districts, depriving them from easy contact with the most vital elements of the city.

TRANSLATION

城市的四大功能

1. 居住

(1) 城市历史核心区的人口密度太大了,就像 19 世纪某些城市外部的工业区一样,达到 1000~1500 人/公顷(合 400~600 人每英亩)。

(2) 在这些拥挤的地区中,生活环境是很糟糕的。其原因包括缺乏足够的用地来安排住宅和绿地,建筑本身也由于开发商投机开发而疏于维护。居民的微薄收入使得他们的灾难更加深重,他们无力采取自我保护措施,死亡率高达 20%。

(3) 城市的扩展不断吞噬着风景优美的绿色周边地带。人们离自然越来越远,公众健康进一步遭到威胁。

(4) 居住建筑布满整个城市,这与公共健康的需求是背道而驰的。

(5) 现实中人烟最稠密的地区往往是最不适于居住的地点,如朝向不好的坡地，易受烟雾和工业气体侵害及易遭水灾的地方。

（6）条件最优越的地区，却往往只安置着最稀疏的人口，这些富人们在这里享受各种便利条件：风和日丽，景色秀美，交通便利而且不受工厂的侵扰。

（7）这种不合理的住宅配置，至今仍然为习惯和名义上公正的城市建筑法规所许可，即分区规划。

（8）沿交通线或者围绕交叉口布置的房屋，因为容易遭受灰尘噪声和尾气的侵扰，不宜作为居住房屋之用。

（9）沿街道两旁安置房屋的传统方式，只能保证少数房屋有充足的日照。

（10）公共建筑也和住宅一样，安排得非常不合理。

（11）尤其是学校，常被设置在交通线上，而且离住宅也太远。

（12）现代城市郊区的发展毫无规划，与城市之间缺乏正常联系。

（13）人们做出各种努力，尝试着把市郊纳入行政管理体系之中。

（14）市郊通常只是一些根本不值得维护的破房陋屋的聚集地。

建议：

（15）从今以后，我们必须把城市中最佳的土地让给居住区，在其布置中充分利用地形之便，并考虑气候、日照、绿地等多种因素。

（16）居住区的选址应充分考虑公众的健康。

（17）必须根据地形特征所限定的居住形态，制定合理的人口密度。

（18）必须保证每套住宅获得最基本的日照时间。

（19）必须禁止住宅沿交通干道布置。

（20）我们应该利用现代技术建造高层建筑。

（21）高层建筑必须保证间距，从而为开阔的绿地留出足够的用地。

2. 休闲

（1）总体而言，目前的开放空间尚不能满足需求。

（2）即便面积足够大，开放空间也常常由于地点不合适而难以服务于广大居民。

(3)城市周边偏远的开放空间不能改善城市内部拥挤的生存条件。

(4)为了方便使用,现有的为数不多的运动设备通常被布置在一些暂时的空地上,这些空地多是将来居住区或工业区的预留地。这说明了这些公共空地时常变动的原因。

(5)周末出游的地点往往不能与城市保持便捷的联系。

建议:

(6)今后任何居住区都必须包括足够的、合理布置的绿色空间,以满足儿童、青年、成年人游戏和运动的需要。

(7)有碍健康的建筑街区必须被拆除,并以绿地代之,从而改善邻近居住区的卫生条件。

(8)新的绿地应有明确的功能,应当包括与住宅紧密联系的幼儿园、学校、少年宫和其他公共设施。

(9)应当创造宜人的周末休闲空间。

(10)应该设置公园、森林、活动场所、露天大型运动场和海滩。

(11)应对现有的自然资源进行评估:包括河流、森林、山丘、山脉、山谷、湖泊和海域。

3. 就业

(1)城市中的工作地点(如工厂、手工车间、商业中心和政府机关等)不再按照理性原则布置。

(2)工作地点与居住地点之间距离过远,联系不便。

(3)通勤时间太长,亟待解决。

(4)由于缺乏对用地及其他要素的预先规划,城市和工业的发展都处于混乱的状态。

(5)城市办公集中在商务区,这些地区占据城市中心最佳的位置,享有最完善的交通系统,自然会成为投机商的掠夺对象。既然这些开发项目都是私人经营的,其自然的发展就缺乏必要的有序性。

(6)必须将工作与居住之间的距离减到最小。

(7)工业区应该独立于居住区,并且它们之间应以绿化带相隔离。

(8)工业区必须靠近铁路、运河或高速公路。

(9)手工业源于城市生活,且与之密不可分,因此必须在城市内部为手工业指定专门的用地。

(10)各种公共或私人运营的城市商业应与居住区、城市内部或附近的工厂和手工作坊保持良好的联系。

4. 交通

(1)现有的城市街道网络是由主干道衍生出来的一套枝状体系。在欧洲,这些干道的建造年代远比中世纪要早,有时甚至可以回溯到远古时代。

(2)今日城市中的主要交通线,最初都是为徒步与马车而设计的,不再能够满足现代机械化交通方式的需要。

(3)道路尺度不当,将严重阻碍未来快速机动交通的运用和城市有序发展的步伐。

(4)道路交叉口之间距离过短。

(5)道路宽度不够,要拓宽这些道路不易操作,而且难以收到显著的成效。

(6)目前的道路似乎失去了控制,在精确性、适应性、多样性和舒适度方面都很差,无法满足现代机动交通的需要。

(7)那些气势磅礴的平面图只讲求形式,却严重阻碍了交通。

(8)当城市需要扩张时,铁路系统往往成为城市化的障碍。铁路包围了居住区,使之孤立,与城市的其他重要部分失去了必要的联系。

VOCABULARY

- population [ˌpɒpjəˈleɪʃən] *n.* 人口;全体居民
- expansion [ɪkˈspænʃən] *n.* (尺寸、数量或重要性的)扩大,增加,扩展
- hectare [ˈhekteər] *n.* 公顷

- aggravate [ˈæɡrəveɪt] vt. 使(局势等)更严重,加剧;使(病情)恶化,加重
- situation [sɪtʃuˈeɪʃən] n. 处境,情况,形势;(尤指城镇、建筑物等的)位置
- extension [ɪkˈstenʃən] n. 伸展;延伸;扩建部分
- sanitary [ˈsænɪtəri] adj. 卫生的,清洁的,干净的
- scattered [ˈskætəd] adj. 分散的;散布的;疏疏落落的
- slope [sləʊp] n. 斜坡;山坡
- zoning [ˈzəʊnɪŋ] n. 分区;分区规划
- crossroad [ˈkrɒsrəʊd] n. 交叉路
- community [kəˈmjuːnəti] n. 社区;群体;社团,团体
- incorporate [ɪnˈkɔːpəreɪt] vt. 包含;将……包括在内
- highrise [haiˈraiz] n. 高层建筑
- insufficient [ɪnsəˈfɪʃənt] adj. 不够的,不足的;不充分的
- downtown [daʊnˈtaʊn] adj. 在市中心(的);朝市中心(的)
- leisure [ˈleʒər] n. 空闲,闲暇,休闲
- slum [slʌm] n. (尤指城市中的)贫民窟,棚户区
- feature [ˈfiːtʃər] n. 特色,特征,特点
- recreation [ˌrekrɪˈeɪʃən] n. 娱乐;消遣(方式);重做;再现
- ameliorate [əˈmiːljəreɪt] vt. 使变好,改善,改进
- complex [kɒmpleks] n. 综合大楼;建筑群
- uncontrolled [ˌʌnkənˈtrəʊld] adj. 不受控制的;失控的
- haphazardly [ˌhæpˈhæzədli] adv. 偶然地,随意地;杂乱地;无序地
- speculation [ˌspekjəˈleɪʃən] n. 猜测;推测;推断
- industrial [ɪnˈdʌstriəl] adj. 工业的,产业的
- fringe [frɪndʒ] n. (地区或群体的)边缘,外围;(活动的)次要部分
- administration [ədˌmɪnɪˈstreɪʃən] n. 管理;经营;行政;(企业、机关

等的)管理人员;行政人员;管理部门
- agglomeration [əˌglɒməˈreɪʃən] n. (不同事物的)聚集;堆积
- antiquity [ænˈtɪkwəti] n. 古物;古董;古迹;古代
- mechanized [ˈmekənaizd] vt. 使机械化
- distance [ˈdɪstəns] n. 距离,路程;冷漠的行为,疏远
- efficiency [ɪˈfɪʃənsɪ] n. 效率;效能;功效
- flexibility [fleksəˈbɪləti] n. 弹性;适应性;灵活性
- barricade [ˈbærɪkeɪd] vt. 设路障于某物的对面、周围或前面 n. 障碍物,路障

EXERCISES

[Choose the right answers]

1. As for low in density developments (middle income dwellings), _____ is not mentioned in the article.

    A. the climate is good

    B. the tax is low

    C. nice view

    D. the transportation is convenient

2. Buildings constructed alongside major routes and around crossroads are unsuitable for dwellings, the reason doesn't include _____.

    A. noise B. high taxes

    C. dust D. noxious gases

3. What is not mentioned in the article about the housing proposal? _____.

    A. Residential areas should occupy the best places

B. Villas and low-rise buildings are prohibited

C. Residential zones should be determined on grounds of health

D. Full use should be made of modern building techniques

4. One question not mentioned in the article about recreation is _____.

A. that it is not readily usable by most of the population

B. outlying open spaces

C. that open spaces are generally insufficient

D. that the open space is too big and wastes the land

5. How many pieces of advice do you have for recreation? _____.

A. 5　　　　　B. 6　　　　　C. 7　　　　　D. 8

6. In the article about work, which of the following description is wrong? _____.

A. Unreasonable arrangement of work and place of residence

B. The transport lines are poorly organized

C. Industry was forced to move out of the city

D. Industry is concentrated in urban centers

7. In the article, the wrong description of traffic is _____.

A. lack of road intersections

B. too many road intersections

C. the inflexible street layout

D. insufficient road width

[Speak your mind]

1. What did you learn from this article?

2. What do you think the recommendations in this article are also useful today?

3. What advice do you have for the city's four functions?

Unit 5   Design With Nature

This chapter was published in its original form as:

Ian L. McHarg (1969), city and countryside, Design With Nature, the Natural History Press, 5.

TEXT

We need nature as much in the city as in the countryside. In order to endure we must maintain the bounty of that great cornucopia which is our inheritance. It is clear that we must look deep to the values which we hold. These must be transformed if we are to reap the bounty and create that fine visage for the home of the brave and the land of the free. We need, not only a better view of man and nature, but a working method by which the least of us can ensure that the product of his works is not more despoliation.

It is not a choice of either the city or the countryside: both are essential, but today it is nature, beleaguered in the country, too scarce in the city which has become precious. I sit at home overlooking the lovely Cresheim Valley, the heart of the city only twenty minutes away, alert to see a deer, familiar with the red-tailed hawk who rules the scene, enamored of the red squirrels, the titmouse and chickadees, the purple finches, nuthatches and cardinals. Yet each year responding to a deeper need, I leave this urban idyll for the remoter lands of lake and forest to be found in northern Canada or the other wilderness of the sea, rocks and beaches where the osprey patrols.

This book is a personal testament to the power and importance of sun, moon, and stars, the changing seasons, seedtime and harvest, clouds, rain and rivers, the oceans and the forests, the creatures and the herbs. They are with us now, co-tenants of the phenomenal universe, participating in that timeless yearning that is evolution, vivid expression of time past, essential partners in survival and with us now involved in the creation of the future.

Our eyes do not divide us from the world, but unite us with it. Let this be known to be true. Let us then abandon the simplicity of separation and give unity its due. Let us abandon the self-mutilation which has been our way and give expression to the potential harmony of man-nature. The world is abundant, we require only a deference born of understanding to fulfill man's promise. Man is that uniquely conscious creature who can perceive and express. He must become the steward of the biosphere. To do this he must design with nature.

TRANSLATION

设计结合自然

我们在城市和在农村一样需要大自然。为使人类能延续下去,我们必须把人类继承的大自然的恩赐保存下来。显然,我们必须对自己拥有的价值有深刻的理解。假如我们要从这恩赐中得益,为勇士们的家园和自由人民的土地创造美好的面貌,我们必须改变价值观。我们不仅需要对人类和自然的关系持有较为正确的观点,而且要有一个较好的工作方法,保证我们中的少数人的工作不会对自然造成更严重的掠夺。

不是说在城市或乡村之间选择何者更重要,而是两者都很重要。但是,今天自然环境在农村也遭到破坏,而在城市中又很稀少,因此,变得十分珍贵。我坐在家里,瞭望着离市中心只有二十分钟路程的美丽的克列谢姆河谷,观看机警的鹿儿,常见到在空中盘旋、领略景色的红尾鹰,还有使人迷恋的红松鼠、长尾山雀、黑头山雀、紫雀和红鸟等。然而,每年我出于更深刻地理解大自然的需要,离开这个城市里的田园景色,到加拿大北部去寻找更偏僻的地方,如自然形成的湖泊和森林或者茫茫大海,多岩石的荒野,鱼鹰游弋的海滩。

本书是关于太阳、月亮、星星、四季变化、播种和收获、云彩、雨水和江湖、海洋和森林、生灵与草木的威力及重要性的个人见解。这些自然要素现在与人类一起,成为宇宙中的同居者,参与到无穷无尽的探索进化的过程中,生动地表达了时光流逝的经过,它们是人类生存的必要伙伴,现在又和我们共同创造世界的未来。

我们不应把人类从世界中分离开来看,而要把人和世界结合起来观察和判断问题。愿人们以此为真理。让我们放弃那些简单化地、割裂地看问题的态度和方法,而给予应有的统一。愿人们放弃已经形成的自我毁灭的习惯,而将人与自然潜在的和谐表现出来。世界是丰富的,满足人类的希望仅仅需要我们理解、尊重自然。人是唯一具有理解能力和表达能力的有意识的生物,因此必须成为生物界的管理员。要做到这一点,设计必须结合自然。

VOCABULARY

- countryside [ˈkʌntrɪsaɪd] *n.* 农村, 乡下; 郊外
- bounty [ˈbaʊntɪ] *n.* 慷慨, 大方; 奖金; 赏金; 大量
- cornucopia [kɔːnjuˈkəʊpɪə] *n.* 丰饶角; 聚宝盆
- inheritance [ɪnˈherɪtəns] *n.* 继承的遗产; 遗传
- visage [ˈvɪzɪdʒ] *n.* 脸庞
- despoliation [dɪˌspəʊlɪˈeɪʃn] *n.* 抢劫; 掠夺
- beleaguer [bɪˈliːgər] *vt.* 围攻; 围
- red-tailed hawk 红尾鹰
- enamor [ɪˈnæmə] *vt.* 使迷恋, 使倾心
- squirrel [ˈskwɪrəl] *n.* 松鼠
- titmouse [ˈtɪtˌmaʊs] *n.* 山雀
- idyll [ˈɪdəl] *n.* (尤指乡下)愉快恬淡的情景(或时期); 田园乐曲; 田园诗
- wilderness [ˈwɪldənəs] *n.* 荒无人烟的地区; 荒野
- osprey [ˈɒspreɪ] *n.* 鹗, 鱼鹰
- patrol [pəˈtrəʊl] *vt.* (尤指士兵或警察)巡逻, 巡查 *n.* 巡逻, 巡查
- testament [ˈtestəmənt] *n.* 遗嘱, 遗言; 证明; 验证
- seedtime [ˈsiːdtaɪm] *n.* 发展阶段; 准备阶段; 播种时期
- herb [hɜːb] *n.* (用作调味品或药材的)香草, 药草
- co-tenants [ˈkəʊˈtenənts] *n.* 共同租地人; 合租人; 共同佃户; 同居者
- participate [pɑːˈtɪsɪpeɪt] *vi.* 参与, 参加
- yearning [ˈjɜːnɪŋ] *n.* 渴望, 切盼, 渴求
- evolution [ˌiːvəˈluːʃn] *n.* 发展; 演变; 演化; 进化

- vivid [ˈvɪvɪd] *adj.* 鲜活的,生动的,(描述、记忆等)栩栩如生的,明亮的
- simplicity [sɪmˈplɪsəti] *n.* 简单,简易;朴素,简朴
- abandon [əˈbændən] *vt.* 放弃;抛弃;遗弃;中止
- unity [ˈjuːnəti] *n.* 联合;一致;团结;和睦
- self-mutilation [ˈselfˌmjuːtiˈleɪʃən] *n.* 自残;自我毁伤
- deference [ˈdefərəns] *n.* 尊重,尊敬
- uniquely [juːˈniːkli] *adv.* 独特地;珍奇地
- perceive [pəˈsiːv] *vt.* 察觉,注意到,意识到;认为;看待
- steward [ˈstjuːəd] *n.* 服务员;管家;组织者,操办者
- biosphere [ˈbaɪəʊsfɪər] *n.* 生物圈

EXERCISES

[Choose the right answers]

1. We have to change our values in order to _____.

A. make cities and villages have the same natural environment

B. get some benefit from the gift of nature

C. make the human race last forever

D. have a more correct view of the relationship between humans and nature

2. The underlined word "despoliation" in the first paragraph refers to _____.

A. complexity

B. destructive effect

C. the act of stripping and taking by force

D. illegality

3. Why does the natural environment become so precious today? _____.

A. Because the natural environment in the city has disappeared

B. Because only remote parts of northern Canada have a natural environment

C. Because the natural environment in the countryside has disappeared

D. Because the natural environment in the countryside has been destroyed to some extent and the urban landscape is very scarce

4. Which statement about natural elements is incorrect? _____.

A. Natural elements are very important to human beings

B. Natural elements are dispensable for humans

C. Natural elements are necessary partners for human survival

D. Natural elements can create the future of the world together with mankind

5. As for the relationship between human and nature, the author hopes that people can _____.

A. take the responsibility of protecting, respecting and understanding nature

B. give up their existing way of life and make people and nature more harmonious

C. keep those simplified and isolated attitudes and methods of looking at problems

D. control nature at will as administrators

[Speak your mind]

1. According to the first paragraph, how can we preserve the gift of nature?

2. In order to understand nature more deeply, what will the author do?

3. What is the book about?

4. According to the last paragraph of the text, what attitude should people hold towards nature?

5. As a student majoring in urban and rural planning, please talk about your understanding of "design with nature".

Unit 6   Accessibility

This chapter was published in its original form as:
Geurs, K. T. end Wee, B. van (2004), Accessibility evaluation of land-use and transport strategies: review end research directions, Journal of Transport Geography. 127-140.

TEXT

Accessibility, a concept used in a number of scientific fields such as transport planning, urban planning and geography, plays an important role in policy making. However, accessibility is often a misunderstood, poorly defined and poorly measured construct. Indeed, finding an operational and theoretically sound concept of accessibility is quite difficult and complex. As a result, land-use and infrastructure policy plans are often evaluated with accessibility measures which are easy to interpret for researchers and policy makers, such as congestion levels or travel speed on the road network, but which have strong methodological disadvantages.

Several authors have written review articles on accessibility measures, often focusing on certain perspectives, such as location accessibility, individual accessibility or economic benefits of accessibility. Our review differs from existing review articles in the following ways.

Firstly, accessibility measures are reviewed from different perspectives, and we do not focus on one specific perspective. The main purpose is to assess the usability of accessibility measures in evaluations of both land-use and transport changes, and related social and economic impacts. Secondly, measures are reviewed according to a broad range of relevant criteria, i. e. (a) theoretical basis, (b) interpretability and communicability, (c) data requirements, (d) usability in social and economic evaluations. This review, based on an extensive literature study, will approach the different perspectives and components of accessibility in Section 2, the accessibility measures in Section 3 and explore the conclusions in Section 4. Future research paths will be outlined in Section 5.

Accessibility is defined and operationalised in several ways, and thus has taken on a variety of meanings. These include such well-known definitions as "the potential of opportunities for interaction" (Hansen, 1959), "the ease with which any land-use activity can be reached from a location using a particular transport system" (Dalvi and Martin, 1976), "the freedom of individuals to decide whether or not to participate in different activities" (Burns, 1979) and "the benefits provided by a transportation/land-use system" (Ben-Akiva and Lerman, 1979). In our study, accessibility measures are seen as indicators for the impact of land-use and transport developments and policy plans on the functioning of the society in general. This means that accessibility should relate to the role of the land-use and transport systems in society, which, in our opinion, will give individuals or groups of individuals the opportunity to participate in activities in different locations. Focusing on passenger transport, we define accessibility as the extent to which land-use and transport systems enable (groups of) individuals to reach activities or destinations by means of a (combination of) transport mode (s). Furthermore, the terms "access" and "accessibility" in the literature are often used indiscriminately. Here, "access" is used when talking about a person's perspective, "accessibility" when using a location's perspective.

A number of components of accessibility can be identified from the different definitions and practical measures of accessibility that are theoretically important in measuring accessibility. Four types of components can be identified: land-use, transportation, temporal and individual.

1. The land-use component reflects the land-use system, consisting of (a) the amount, quality and spatial distribution opportunities supplied at each destination (jobs, shops, health, social and recreational facilities, etc.), (b) the demand for these opportunities at origin locations (e. g. where inhabitants live), (c) the confrontation of supply of and demand for opportunities, which may result in competition for activities with restricted capacity such as job and school vacancies and hospital beds.

2. The transportation component describes the transport system, expressed as the disutility for an individual to cover the distance between an origin and a destination using a specific transport mode, included are the amount of time (travel, waiting and parking), costs (fixed and variable) and effort (including reliability, level of comfort, accident risk, etc.). This disutility results from the confrontation between supply and demand. The supply of infrastructure includes its location and characteristics (e. g. maximum travel speed, number of lanes, public transport timetables, travel costs). The demand relates to both passenger and freight travel.

3. The temporal component reflects the temporal constraints, i. e. the

availability of opportunities at different times of the day, and the time available for individuals to participate in certain activities (e. g. work, recreation).

4. The individual component reflects the needs (depending on age, income, educational level, household situation, etc.), abilities (depending on people's physical condition, availability of travel modes, etc.) and opportunities (depending on people's income, travel budget, educational level, etc.) of individuals. These characteristics influence a person's level of access to transport modes (e. g. being able to drive and borrow/use a car) and spatially distributed opportunities (e. g. have the skills or education to qualify for jobs near their residential area), and may strongly influence the total aggregate accessibility result. Several studies (e. g. Cervero et al. ,1997; Shen,1998; Geurs and Ritsema van Eck,2003) have shown that in the case of job accessibility, inclusion of occupational matching strongly affects the resulting accessibility indicators.

The picture below shows the relationships between these components of accessibility (as defined above), and relationships between the components themselves: here, the land-use component (distribution of activities) is an important factor determining travel demand (transport component) and may also introduce time restrictions (temporal component) and influence people's opportunities (individual component). The individual component interacts with all other components: a person's needs and abilities that influence the time, cost and effort of movement, types of relevant activities and the times in which one engages in specific activities. Furthermore, accessibility may also influence the components through feedback mechanisms, i. e. accessibility as a location factor for inhabitants and firms (relationship with land-use component) influences travel demand (transport component), people's economic and social opportunities (individual component) and the time needed to carry out activities (temporal component).

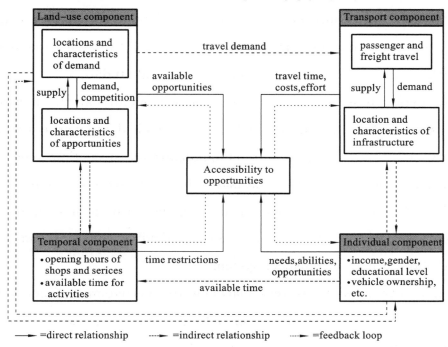

K.T.GEURS,B.van Wee I Journal of Transport Geography 12(2004)127-140

Relationships between components of accessibility

Following our definition of accessibility, an accessibility measure should ideally take all components and elements within these components into account. In practice, applied accessibility measures focus on one or more components of accessibility, depending on the perspective taken. Four basic perspectives on measuring accessibility can be identified.

1. Infrastructure-based measures, analysing the (observed or simulated) performance or service level of transport infrastructure, such as "level of congestion" and "average travel speed on the road network". This measure type is typically used in transport planning.

2. Location-based measures, analysing accessibility at locations, typically on a macro-level. The measures describe the level of accessibility to spatially distributed activities, such as "the number of jobs within 30 min travel time from origin locations". More complex location-based measures explicitly incorporate capacity restrictions of supplied activity characteristics to include competition effects. Location-based measures are typically used in urban planning and geographical studies.

3. Person-based measures, analysing accessibility at the individual level, such as "the activities in which an individual can participate at a given time". This type of measure is founded in the space-time geography of Hagerstrand (1970) that measures limitations on an individual's freedom of action in the environment, i.e. the location and duration of mandatory activities, the time budgets for flexible activities and travel speed allowed by the transport system.

4. Utility-based measures, analysing the (economic) benefits that people derive from access to the spatially distributed activities. This type of measure has its origin in economic studies.

TRANSLATION

可 达 性

可达性是一个在交通规划、城市规划、地理学等很多科学领域使用的概念,在政策制定方面扮演着重要角色。然而,可达性是一个经常被误解、不容易定义且不易度量的概念。事实上,很难找到一个可用且有理有据的可达性概念。因此,土地规划和基础设施政策规划经常采用容易被研究者和决策者理解的可达性测度方法来进行评估,例如路网的拥挤程度或者交通速度,但这种方法在方法论上有很大的缺点。

一些作者曾写过可达性测度方法的综述,经常聚焦于一些特定方面,例如区位可达性、个人可达性或者可达性的商业收益。我们的综述在以下方面不同于已有的综述论文。首先,从不同视角评估可达性,我们没有只关注于某一特定视角。主要目的是实现可达性测量的可用性,在土地使用和交通变化双重变革,以及其他社会经济方面影响的背景下。其次,根据一系列的标准来进行方法综述,例如(a)理论基础,(b)可解释性和可交流性,(c)数据要求,(d)社会经济评价的有效性。这个基于广泛文献研究的综述,将在第二部分研究可达性的不同视角和组成,第三部分是可达性评价,第四部分是结论研究,第五部分概述未来的研究途径。

可达性是通过多种方式来定义和操作的,因此其具有多种含义。其中包括著名的概念,例如互动机会的潜力(Hansen,1959),土地开发行为通过一个特别的交通系统到达一个地点的容易程度(Dalvi and Martin,1976),个人决定是否参加不同活动的自由(Burns,1979),土地和交通系统的效用(Ben-Akiva and Lerman,1979)。在我们的研究中,可达性评价被认为是土地利用、交通发展、社会公共政策计划的指示灯。这意味着可达性应该与社会中土地开发和交通系统相关联,这将给予个人或团体去参与不同地方的活动的机会。聚焦客运,我们认为可达性是土地利用和交通系统使人通过运输方式到达活动区或目的地的程度。此外,文献中的"到达"和"可达性"这两个词经常被不加区分地使用。在这里,当以一个人的视角来讨论时,使用"到达",而当我们使用一个地点的视角时,使用"可达性"。

可达性的许多组成部分可以从可访问性的不同定义和实际度量中识别出来,这些定义和度量在理论上对可达性评价很重要。四种类型的组成成分已经可以被识别:土地使用、交通、时间和个人。

1.土地使用要素反映土地使用系统,土地使用系统由以下部分组成:(a)每个目的地蕴含的机会的数量、质量和空间分布(就业、商业、健康、社会和娱乐设施等），(b)在最初位置（例如居住地）产生的对这些机会的需求，(c)机会供给和需求的冲突，可能导致对有限资源的竞争行为，例如就业岗位、学校招生数量和医院床位等。

2.交通要素反映交通系统，表示为个人使用某类交通模式完成从起点到终点距离的负效用，包括时间（移动、等候、停泊），支出（固定和可变）和努力（包括可靠程度、舒适水平和事故风险等）。这些负效用产生的原因在于供求之间的矛盾。基础设施的供给包括了它的区位和特征（例如最大交通速度，车道数量，公共交通时刻表，交通花费）。需求与旅客和货运两者都有关联。

3.时间要素反映了时间的强制性，也就是在每天不同时段的机会的可用性，以及某人参加某些活动（例如工作、娱乐）的可用时间。

4.个人因素反映个人的需求（取决于年龄、收入、受教育程度、家庭情况等）、能力（取决于个人身体条件、可用的出行方式等）和个人机遇（取决于收入、交通预算、受教育程度等）。这些特质影响了一个人的出行方式的水平（例如有能力去驾驶汽车和租用汽车）、空间分布机会（例如拥有一定的技能和受教育程度并在他们居住地附近得到就业的资格），可能对整个可达性结果造成巨大影响。一些研究（例如 Cervero et al.，1997；Shen，1998；Geurs and Ritsema van Eck，2003)指出就工作可达性而言，职业匹配内容对可达性的结果有很大的影响。

下图（见前文图）表明了这些因素和可达性（上面所定义的）之间的关联，以及这些因素本身之间的关联：这里，土地利用因素（行为分布）是决定交通需求的一个重要因素（交通因素），可能也会引出时间限制（时间因素）和影响人们自身的机会（个人因素）。个人因素和其他所有因素相互作用：个人的需求和能力影响了时间、支出和行动消耗，以及相关的活动类型和特殊活动所投入的时间。此外，可达性可能通过回馈机制来影响这些因素，也就是说可达性作为个人或者公司的区位要素（土地利用因素方面的关联）影响了交通需求（交通要素），人的经济状况和社会机遇（个人因素），以及付诸行为所需的时间(时间因素)。

根据我们对可达性的定义,可达性评价应该完美地包含所有因素,把所有因素综合在原理之中。在实践中,应用可达性评价集中于一个或者多个可达性因素,取决于它所采取的角度。已确定可达性评价的四个基本角度如下。

1. 基于基础设施的测量,分析(观察或模拟)交通基础设施的表现或服务水平,例如路网的拥挤水平和平均速度。这类测量通常被用于交通规划。

2. 区位测量,分析区位的可达性,典型是在宏观层面。这些措施描述了空间分布式活动的可达性水平,例如从原点出发半小时内的就业岗位数量。更多复杂的区位测量明确包含所提供的活动类型的潜力限度,从而被纳入竞争效应。区位测量经常被用于城市规划和地理研究。

3. 以人为基础的测量从个人层面分析可达性,例如个人在给定时间内可以参加的活动。这种测量是建立在哈格斯特朗(1970)的时空地理学上的,测量人在环境中活动的自由程度,也就是强制性活动的地点和持续时间,自由活动的时间成本和交通系统中所允许的行进速度。

4. 效用测量,分析人们从空间活动中所获得的(经济)收益。这种测量方法源自经济学研究。

VOCABULARY

- accessibility [əkˌsesəˈbɪləti] n. 可达性
- network [ˈnetwɜːk] n. 网络;广播网;网状物
- congestion [kənˈdʒestʃən] n. 拥挤;拥塞;淤血
- infrastructure [ˈɪnfrəˌstrʌktʃər] n. 基础设施;公共建设
- evaluation [ɪˌvæljuˈeɪʃn] n. 评价;评估;估价;求值

- perspective [pə'spektɪv] n. 远景；透视图；观点
- requirement [rɪ'kwaɪəmənt] n. 要求；必要条件；必需品
- outline ['aʊtlaɪn] vt. 概述；略述
- definition [ˌdefɪ'nɪʃən] n. 定义；清晰度；解说
- ease [iːz] n. 轻松，舒适；安逸，悠闲
- interaction [ˌɪntə'rækʃən] n. 相互作用；交互作用；互动
- indicator ['ɪndɪkeɪtər] n. 指标，标志，迹象
- destination [ˌdestɪ'neɪʃən] n. 目的地，终点
- vacancy ['veɪkənsi] n. 空缺；空位；空白；空虚
- inhabitant [ɪn'hæbɪtənt] n. 居民；居住者
- disutility [ˌdɪsjuː'tɪləti] n. 无用，无效；负效用
- confrontation [ˌkɒnfrʌn'teɪʃən] n. 对抗；面对；对峙
- freight [freɪt] n. 货运；运费；船货
- constraint [kən'streɪnt] n. 约束；局促，态度不自然；强制
- temporal ['tempərəl] adj. 时间的
- inclusion [ɪn'kluːʒən] n. 包含；内含物
- budget ['bʌdʒɪt] n. 预算，预算费
- definition [ˌdefə'nɪʃən] n. 定义
- simulated ['sɪmjəleɪtɪd] adj. 模拟的；模仿的；仿造的
- space-time [ˌspeɪs'taɪm] n. 时空
- matrix ['meɪtrɪks] n. 矩阵；模型

EXERCISES

[Choose the right answers]

1. In the article, what is not mentioned about accessibility? _____.

A. Accessibility is a poorly defined and poorly measured construct

B. Accessibility is a well-defined concept

C. Accessibility has a very important role in policy making

D. Accessibility is a concept used in a number of scientific fields

2. In the author's review of the accessibility, what's the misdescription? _____.

A. This review is based on an extensive literature study

B. They focus on one specific perspective

C. There are extensive standard reviews

D. Accessibility measures are reviewed from different perspectives

3. What's the wrong description of the four components of accessibility? _____.

A. Residential areas should occupy the best places

B. The transportation components include time, costs and effort

C. The individual component reflects the needs, abilities and opportunities

D. The time element reflects the non-compulsion of time

4. What is not mentioned in the description of Fig. 1? _____.

A. Accessibility affects individuals, transportation and time

B. The land-use component determines travel demand

C. This figure shows the relationships between these components and accessibility

D. Accessibility does not affect these factors through feedback mechanisms

5. How many basic angles are there to evaluate accessibility? _____.

A. 2   B. 4   C. 6   D. 8

[Speak your mind]

1. What did you learn from this article?

2. Can you describe accessibility in your own words? Write it down.

3. How many components of accessibility are there? What are they? Give your opinion.

Unit 7   Ecological Planning Methods

This chapter was published in its original form as:
D. Palazzo and F. Steiner(2011), Urban Ecological Design: A Process for Regenerative Places, Knowledge, Island Press, 55-56.

TEXT

Ecological planning emerged in North America during the 1960s. The pioneering studies of McHarg and others, including George Angus Hills in Canada and Philip H. Lewis in Wisconsin, became well known among professionals, scholars, and students. These studies and applications, which originated from many precedents in regional planning and landscape architecture, have evolved rapidly since the 1980s.

In *The Lining Landscape*, Steiner presented an ecological planning model composed of eleven steps. The third and fourth steps are, respectively, dedicated to regional and local level inventory and analysis. Steiner's book provides a list of elements, from regional to local, to he inventoried in the design process.

* Regional climate—temperature and precipitation

* Geology—geological maps to evaluate the suitability of an area as a construction site

* Terrain—physiography (elevation and slope)

* Water—water budget (precipitation, uses, and groundwater), hydrological cycle, flooding areas, water quality, hydrologic system, water supply, and sewage treatment systems

* Soils—characteristics, soil survey, and soil capability classification (the survey particularly helps with understanding land uses and land values for specific activities)

* Microclimate—ventilation, solar radiation, albedo, and temperatures

* Vegetation—plant communities; rare, endangered, or threatened plants; native and disturbance adaptive plants

* wildlife—species; habitat values; habitat of rare, endangered, or threatened species

* Existing land use and land users—the physical arrangement of space utilized by humans, ownership (public and private), settlement patterns, buildings, and open space types

In *The Granite Garden*, Anne Spirn advances McHarg's method for urban areas. She uses the clever subtitle "Urban Nature and Human Design" to explain her approach. Spirn illustrates the importance of understanding the natural settings of cities, including their air, geology, water, plants, and animals, to create better urban spaces.

TRANSLATION

生态规划方法

20世纪60年代,北美出现了生态规划。麦克哈格等人的开拓性研究在专业人士、学者和学生中变得闻名,这些人还包括加拿大的乔治·安格斯·希尔斯和威斯康星州的菲利普 H. 刘易斯。这些源于区域规划和景观建筑的研究和应用自20世纪80年代以来取得了迅速发展。

在《衬内景观》一书中,斯坦纳提出了一个由十一个步骤组成的生态规划模型。其中第三步和第四步分别是致力于区域和地方层面的盘点和分析。斯坦纳的书中列出了一系列从地域到地方的元素,以及他在设计过程中列出的元素。

* 区域气候——温度和降水
* 地理学——地质地图,用于评估某个区域作为施工场地的适用性
* 地形——地形地貌(海拔和坡度)
* 水资源——水预算(降水、使用和地下水),水文循环,洪水地区,水质,水文系统,供水和污水处理系统
* 土壤——土壤特性、土壤调查和土壤肥力分类(该调查特别有助于了解特定活动的土地使用和土地价值)
* 微气候——通风、太阳辐射、反射率和温度
* 植被——植物群落;稀有、濒危或受威胁的植物物种;原生和干扰适应性植物
* 野生动物——物种;栖息地价值;稀有、濒危或受威胁物种的栖息地
* 现有土地使用和土地使用者——被人类利用的符合自然规律的空

间布局、所有权(公共和私有)、定居模式、建筑和开敞空间类型

在《花岗岩花园》一书中,安妮·斯本推进了麦克哈格在城市地区的方法。她用巧妙的副标题"城市自然与人类设计"来解释她的方法。斯本阐明了理解城市自然环境的重要性,包括空气、地质、水、植物和动物,以创造更优质的城市空间。

VOCABULARY

- ecological [ˌiːkəˈlɒdʒɪkəl] *adj.* 生态的,生态学的
- emerge [ɪˈmɜːdʒ] *vi.* 浮现;摆脱;暴露
- pioneering [ˌpaɪəˈnɪərɪŋ] *adj.* 首创的;先驱的
- application [ˌæplɪˈkeɪʃən] *n.* 应用;申请;应用程序
- precedent [ˈpresɪdənt] *n.* 先例;前例
- regional [ˈriːdʒənl] *adj.* 地区的;局部的;整个地区的
- landscape [ˈlændskeɪp] *n.* 风景;风景画;景色;山水画;乡村风景画;地形
- architecture [ˈɑːkɪtektʃər] *n.* 建筑学;建筑风格;建筑式样;架构
- evolve [ɪˈvɒlv] *vt.* 发展;进化;使逐步形成;推断出
- inventory [ˈɪnvəntəri] *n.* 存货,存货清单;详细目录;财产清册
- precipitation [prɪˌsɪpɪˈteɪʃən] *n.* 降水(冰雹);坠落;鲁莽;沉淀,沉淀物
- geology [dʒiˈɒlədʒi] *n.* 地质学;地质情况
- geological [ˌdʒiəˈlɒdʒɪkəl] *adj.* 地质的,地质学的
- construction [kənˈstrʌkʃən] *n.* 建设;建筑物;解释;造句
- terrain [təˈreɪn] *n.* 地形,地势;领域;地带
- physiography [ˌfɪziˈɒgrəfi] *n.* 地文学,地相学;自然地理学;自然

现象志

- elevation [ˌelɪˈveɪʃən] n. 高地；海拔；提高；崇高；正面图
- slope [sləʊp] n. 斜坡；倾斜；斜率
- hydrologic [ˌhaɪdrəˈlɑdʒɪk] adj. 水文的
- sewage [ˈsuːɪdʒ] n. 污水；下水道；污物
- microclimate [ˈmaɪkrəʊˌklaɪmət] n. 小气候，微气候
- ventilation [ˌventəˈleɪʃən] n. 通风设备；空气流通
- radiation [ˌreɪdiˈeɪʃən] n. 辐射；放射物
- albedo [ælˈbiːdəʊ] n. （行星等的）反射率；星体反照率
- native [ˈneɪtɪv] adj. 本国的；土著的；天然的；与生俱来的；天赋的
- disturbance [dɪˈstɜːbəns] n. 干扰；骚乱；忧虑
- wildlife [ˈwaɪldlaɪf] n. 野生动植物
- species [ˈspiːʃiːz] n. 物种；种类
- threaten [ˈθretən] v. 威胁；恐吓
- utilize [ˈjuːtəlaɪz] vt. 利用
- ownership [ˈəʊnəʃɪp] n. 所有权；物主身份
- settlement [ˈsetəlmənt] n. 解决，处理；结算；沉降；殖民
- advance [ədˈvɑːns] vt. 提出；预付；使……前进；将……提前

EXERCISES

[Choose the right answers]

1. Why did pioneering studies by McHarg and others become famous among professionals, scholars, and students? _____.

A. Because these studies and applications of regional planning and landscape architecture have developed rapidly since the 1980s

B. Because these pioneering studies of McHarg and others have been sought after by George Angus Hills and Philip H. Lewis

C. Because McHarg has great personal influence

D. Because these studies of McHarg and others have made great contributions to urban development

2. What Steiner mentioned in his book *The Lining Landscape*, except _____.

A. an ecological planning model composed of eleven steps

B. regional and local level inventory and analysis

C. an urban planning model composed of eleven steps

D. a series of elements from region to place

3. What are the elements listed by Steiner during his design process?

A. Regional climate, terrain, microclimate, universe, wildlife

B. Regional climate, geography, microclimate, water resources, vegetation

C. Terrain, human activities, water resources, vegetation, wildlife

D. Terrain, soils, earth, microclimate, wildlife

4. Anne Spence spelled out _____ in the book Granite Garden.

A. the importance of understanding the natural settings of cities

B. the relationship between air, geology, water, plants and animals

C. that urban natural environment is more important than human design

D. that only natural environment can create better urban space

5. When did ecological planning come into being? _____.

A. Before the 1960s

B. During the 1960s

C. During the 1970s

D. During the 1980s

[Speak your mind]

1. How do you understand "ecological planning"?

2. What elements are included in the design process according to Steiner's ecological planning model?

3. What does microclimate mean in the passage?

4. In the book of *The Granite Garden*, how does Anne Spirn provide to explain her ecological planning approach?

5. What can we do to create a better urban space?

Unit 8   Patrick Geddes' Visual Thinking

This chapter was published in its original form as:

Mercedes Diaz (2017), Patrick Geddes' visual thinking, in EGA Revista de Expression Grafica Arquitectonica 22(29):256. DOI: 10.4995/ega.2017.7374

TEXT

Patrick Geddes' most known image is undoubtedly the Valley Section, usually described as an illustration of his idea of region. Although this is the most widely spread image, in truth different kinds of graphic documents, as well as other proposals related to the visual, acquire an outstanding prominence in his work.

The Valley Section illustrates Geddes' concept of region. There are several versions of it, the first from 1909. Besides the Section as a graphic image we find this concept mentioned in some texts, the lengthiest being from 1923.

It shows a longitudinal section along the course of a river, from its source in the mountains to its broad mouth in the sea. The river valley is proposed as a typical basic unit for the study of the region, which encompasses natural or environmental conditions represented in the drawing through plants, with occupations or basic activities represented by tools, and with types of settlement represented by their outlines.

Geddes depicts his model as a succession going downwards from natural woods, breeding lands, through the fertile valley to the sea, with their matching natural occupations and typical settlements, from the small village to the big industrial and port city through the small mountain village on the main track, the small mountainside town with its own marketplace, or the prosperous market town at the center of the valley, crossroads on the ways to other mountain villages. The region would reflect as well different evolutive stages characterized by different ways of subsistence, all of them taking place in the region with its succession of occupations: hunter, woodman, miner, shepherd, peasant...

As for the image itself the following features stand out.

① It contains descriptive and symbolic elements displayed in a diagrammatic way.

②It expresses a complex concept where different aspects take part, as well as the idea of evolution in time.

③It is seemingly simple despite the complexity of its message.

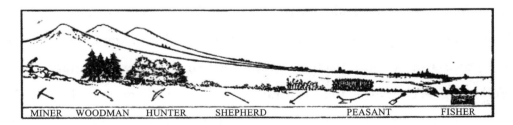

A certain relationship can be found between this image and the peculiar style of his writings, where the message is often disguised as a drama by the means of small stories. In his text The Valley Section from

1923 we find an example of this as, describing the occupations, he alludes to historical personages or biblical passages affirming that "simple and vivid interpretations such as this appear over the whole range of occupations from mountain to sea and come together to develop a re-understanding of history from the evolutionary standpoint". In both instances, image and text, the same resource is employed: the representation of the city in the region as a drama evolved in space and time through known stories or situations which imply additional meanings. This image has sometimes been defined as a diagram, but although indeed it shows a certain diagrammatic character, we find that what really characterizes it is the fact that it conveys a message that goes further than what is seen, through the representation of a drama showing the relation between man and environment, actor and scene we might say, through action. This is why we consider more appropriate a definition of it as a symbolic narrative, a graphic narrative in this instance that finds its parallel in the written narrative.

The notation of life is the visual representation of a theory of the city. Geddes produces in 1905 a first version of his diagram where the basic features as to his idea of the city are already developed. In 1927 he publishes a more complex version, in which the idea of the city underlies the process of action-reflection through which the city is accomplished. Following Welter, the four central terms of the diagram published in 1927, "Town", "School", "Cloister" and "City in deed" represent the so-called Town-City formula, which encompasses four evolutive levels within the city. On the margins of the diagram the terms "Acts", "Facts", "Thoughts" and "Deeds" constitute the Act-Deed formula, which encompasses four parallel levels of action-reflection as levers of the transforming process.

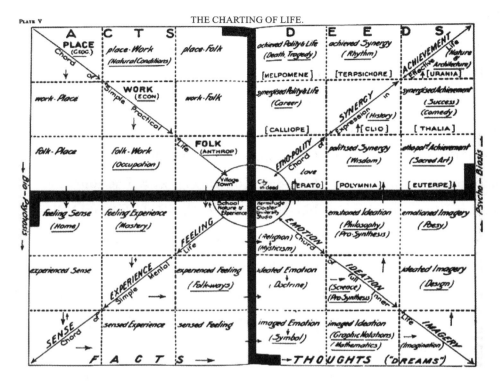

The formula Town-City represents as a whole what Geddes calls "the city complied". On a first level would find place the city of daily action, the city which results from the interaction place-work-folk. It would correlate to the outer objective world (Town). The following two levels show cultural products on two successive steps, the first being that of generalized knowledge and norms (School) as well as research and the second being that of production by the intellectual elite (Cloister), entrusted in Geddes' words with the elaboration of "a synthesis of a new kind" which is classified in three categories: ideas, ideals and aesthetics. Last, the fourth level would be the one where the new synthesis is accomplished in the city through its reflection on culture, politics and art.

In the equally gradual process described in the formula Act-Deed, the first level (Act) would be that of individual action or of social behaviour in daily life, once again defined as relation between place-work-folk (Chord of Simple Practical Life). The two following levels mean two successive degrees of mental internalization from the objective world, the first (Facts) through the terms feeling-experience-sense (Chord of Simple Mental Life), the second (Thoughts) through the respective terms of emotion-ideation-imagery (Chord of Full Inner Life). The fourth and last level is again one of action like the first and means the accomplishment of ideas through the concepts ethnopolity-synergy-achievement (Chord of Expression in Effective Life).

As an image in itself the following characteristics can be singled out.

① Its graphic content is completely textual, displayed in a diagrammatic way.

② It shows relations at different levels, within the same evolutive stage and between different stages.

③ It expresses an idea of evolution through stages which are successive and yet simultaneously present.

On the occasion of the publication of the first version of the scheme, Geddes advocates the use of an "abstract method for notation and for interpretation" for the study of cities, as others sciences do it, using the expression "thinking machines".

They also carry on their work by help of definite and orderly technical methods, descriptive and comparative, analytic and synthetic. These, as far as possible, have to be crystallised beyond their mere verbal statement into formulae, into tabular and graphic presentments, and thus not only acquire greater clearness of statement, but become more and more active agencies of inquiry—in fact, become literal "thinking-machines".

In this double function as graphic representation and means of inquiry lies the quality of what we understand as diagram and Geddes calls thinking machine. The notation of life would then be a means of putting in order and linking multiple concepts, the representation of an outcome for the sake of its communication which is made easy by its visualization. A diagram derived in this case exclusively from a textual content.

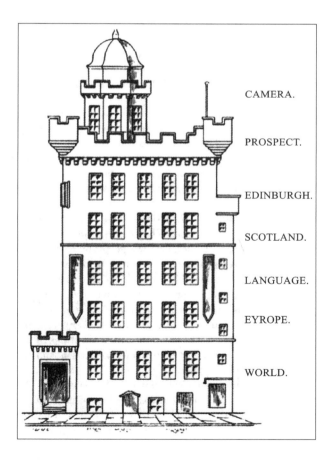

The Outlook Tower is described by Geddes as a civic observatory and laboratory, a center from which the study and knowledge of the city in the region can be fostered. It was "a tall old building, high upon the ridge of Old Edinburgh, it overlooks the city and even great part of its region". Geddes acquires in 1892 this six-storey building which is provided with a camera obscura, he transforms it in his famous Outlook Tower. Along with civic exhibitions, it is the most outstanding proposal in his idea of civism. Visits would begin at the upper level, with a straight view from its roof and through the projection in the camera obscura, which according to Geddes offered impressions of aesthetic or artistic value, the outlook of the artist. From this point, there came in succession in a downward itinerary.

Here, on occasion, is set forth the analysis of the outlook in its various aspects astronomic and topographical, geological and meteorological, botanical and zoological, anthropological and archeeologic, historical and economic, and so on. The storey below this prospect is devoted to the City. Its relief-model maps, geological and other, are here shown in relation to its aspects and beauty expressed in paintings, drawings, photographs, etc.; while within this setting there has been gradually prepared a Survey of Edinburgh, from its prehistoric, origins, and throughout its different phases, up to the photographic details of the present day.

The following levels are devoted to Scotland, Great Britain and the English-speaking world, Europe or the West, the East and Man.

Among the different functions fulfilled by the Tower it is interesting to regard it as: ① Observatory, related to the straight perception of reality; ② Camera obscura, with projection as the first manipulation of the real; ③ Spatial analogy of the process knowledge-action concerning the city through successive steps: perception, introspection (the outlook of the artist), analysis (prospect of the sciences), synthesis and spreading.

The most characteristic among the contents displayed by the Tower would be the straight view of the region of Edinburgh. In several moments the author remarks the importance of general sights on the concrete, of the concrete as the bearer of the sought-after synthesis of knowledge, in what he calls the synoptic view.

What if the long-dreamed synthesis of knowledge, which thinkers have commonly sought so much in the abstract, be really more directly manifest around us, in and along with our surveys of the concrete world? What if Aristotle, that old master of knowledge, turns out to have been literally, and not merely metaphorically, speaking in urging "the synoptic vision"? For surely "general views" may well be helped by general views.

A further reading of the Tower as a spatial analogy of the process knowledge-action would be possible. The last sequence in this process that we have described as perception-introspection-analysis-synthesis-spreading would be symbolized by the exit to the street, to action in real life fuelled by the knowledge that has been acquired, so that the sequence would be not linear but round, or rather spiral, being this a symbol repeatedly employed by Geddes in his schemes.

The examples analysed have hopefully contributed to the elucidation of the meaning of Geddes' so-called visual thinking. Further examples which deserved Geddes' attention could be added. To mention a few: the demographic map of London, upon which he bases his idea of agglomeration, the maps drawn by Ch. Booth in his analysis of London, with "its observant study of streets and quarters, and of the conditions and life of their inhabitants"; or the numerous images collected or made on purpose in order to be shown in the civic exhibitions which he

organised. A huge repertoire of proposals concerning the visual as a means of perception, analysis, reasoning and communication of ideas and theories, which despite its diversity or even disparity shows coherence within a way of thinking whose principles would be the following.

①Reality as a starting point, object of analysis and of transforming action.

②The importance of general views, as much in their scope, spatial or temporal, as in their approach, multidisciplinary.

③ The importance of education as transmission of acquired knowledge to the whole society, main lever for the improvement of reality.

The concern for the understanding of the process knowledge-action would stand as the basis of all this, with the visual image being the key tool.

The reality that Geddes knew is undoubtedly different from that of today, but the need to continue thinking about the nature of the city and about the mechanisms and tools for its improvement is the same. It is here where, in our opinion, lies the importance of his legacy.

TRANSLATION

帕特里克·盖迪斯的视觉思维

帕特里克·盖迪斯最著名的图示无疑是山谷截面图,通常被描述为他对区域这一概念的诠释。虽然山谷截面图是最为广泛传播的,但实际上盖迪斯的成果中,其他类型的图形文档,以及其他与视觉相关的提议,都取得了杰出的成就。

山谷截面图说明了盖迪斯的区域概念。有好几个版本,最初的是在 1909 年。除了从图像部分,我们还在一些文本中发现了这个曾被提到的概念,最初的是在 1923 年。

图片显示了沿河的纵向剖面,其剖面从山脉源头直到宽阔的海口。河谷被提议作为研究该区域的典型基本单位,包括通过植物绘制来表现的自然及其环境,以工具来描述职业或者基本活动,以它们的轮廓来描述其群落的类型。

盖迪斯将他的模型描述为一个连续向下的模型,从自然森林到繁育的大地,穿过肥沃的山谷到达大海,包括与其匹配的自然职业和典型的定居点;从山谷小镇穿过山间小镇的主要轨道到达大型的工业或港口城市,山间小镇拥有自己的市场,或者将繁荣的市场安放在山谷的中心或者通往其他村庄的十字路口上。该地区也将以不同的生存方式为特征来反映不同的发展阶段,所有这些阶段都发生在该地区及其一系列职业,包含猎人、樵夫、矿工、牧羊人、农民……

至于图像本身,有以下特点。

①它包含描述性和符号化的元素,以图表的方式显示。

②他表达了一个复杂的概念,其涉及不同的方面,以及时间演变的概念。

③它看起来很简单,尽管其包含了复杂的信息。

这个图像与他独特的写作风格之间存在着一定的联系,信息所在往往通过小故事的形式伪装成戏剧。从他的关于1923年的山谷截面图的文本部分,我们发现了这样一个例子,在描述职业过程中,他暗指历史人物和《圣经》章节并断言道:"这种简单而生动的解释出现在从山脉到海洋的整个职业范围内,并且一起以发展的角度再认识历史。"在图像和文字两种情况下,都使用了相同的资源:该地区的城市表现为一个通过已知的有着特殊意义的故事或情境在空间和时间内逐步形成的戏剧。这个图像有时候会被做成一个特定的图表,尽管他确实展现了一个图解的特性,但事实上是我们发现他所真正描绘的内容传达出一个信息,即看得更远,通过戏剧的形式,即演员通过在场景中的活动来展示人与环境之间的关系。这就是为什么我们认为他将信息定义为一个象征故事显得更恰当,在这种情况下,图表叙事和文本叙事有相同之处。

生命的符号是城市理论的视觉表现。1905年,盖迪斯完成了他的图表的第一个版本,其中关于他对城市的想法的基本特征已经得到了发展。1927年,他发行了一个更复杂的图表,在这个版本中,城市思想成了行动反思过程的基础,通过这个过程,城市才得以实现。继韦尔特之后,在1927年出版的图表中有四条中心思想,城镇,学校,修道院和城市契约代表了所谓的城市准则,其包含了城市的四种发展水平。在图表的边缘,术语——行动、事实、思想、契约构成了有行为约束力的契约,其包含了四个平行层面的行为反思,其作为转变过程中的杠杆。

城镇-城市这个公式代表了盖迪斯所说的城市服从。在第一个层面上,将会发现城市日常活动地点,城市的结果来自地方工作人员的互动。其将与外部客观世界(城市)相关联。下面的两个层面将用两个连续的步骤来展示文化产品,第一个是广义上的知识、规范(学校)以及研究,第二个是智慧的精英的作品(修道院),其被盖迪斯详细解释为一种新型的综合体。它被分为三种类别:思想,理想和美学。最后,第四层面是通过文化、艺术和政治的反思,在城市里建设新的综合体。

在可约束行为契约准则的描述下,同样渐进的过程,第一层次(行动)是日常生活中的个人行为和社会行为,再次定义地点-工作-人之间的关系(简单实用生活的和弦)。以下两个层次是指来自客观世界的两个连续的心理内化程度,第一个层次(事实)通过感觉-体验-意识(简单心理生活的和弦),第二个层次(思想)通过各自的情感-想象-意象(完整内心生活的和弦)。第四和最后一个层次也是和第一层次一样的活动,意味着通过"民族政策-协同-成就"这一概念实现理想(有效生活中表达的和弦)。

作为一种形象本身,可以突出以下特征。

①它的图形内容完全是文本内容,以图解的方式显示。

②它表现出不同层次的关系,在同一发展阶段和不同阶段之间。

③它表达了一种进化的思想,它经历了一个连续的但同时又存在的阶段。

在第一版方案出版的时候,盖迪斯主张使用一种"抽象的记数和解释方法"来研究城市,就像其他的科学那样,使用"思考机器"这种表达方式。

他们还通过明确而有序的技术方法、描述性和比较性、分析性和综合性来进行他们的工作。这些必须尽可能地具体化,从单纯的口头陈述变为规则,列出表格和图表演示,不仅获得更清晰的陈述,而且成为越来越积极的查询机构——事实上,成为字面上的"思考机器"。

在这种双重功能中,图形表示和查询方法就是我们所理解的图表和盖迪斯称为思考机器的东西的质量。之后,生命的符号将是一种把多种概念有序排列并联系在一起的方法,结果表明为了沟通的目的,使其形象简单化。在这种情况下,图表仅仅只是从文本所派生出来的。

瞭望塔被盖迪斯描述为一个市民天文台和实验室,在该中心可以进行对该地区城市的研究和知识获取。它是"一座高大的古老建筑,耸立在老爱丁堡的山脊上,可以俯瞰整个城市,甚至当地大部分地区"。1892年,盖迪斯获得了这幢六层楼的建筑,它配有一个摄影暗室,他在他著名的瞭望塔里改造了它。与市民展览会一样，这也是他公民精神思想中最突出的提议。参观将从上层开始，从屋顶上直接观看，通过摄影暗室的投影，根据格迪斯的说法，摄影暗室为艺术家提供了审美或艺术价值的印象。从这里开始，就有了一个继续向下的过程。

在这里，我们不时地从天文学和地形学、地质学和气象学、植物学和动物学、人类学和考古学、历史学和经济学等各个方面来进行前景分析以及阐述展望。这个展望之下的楼层是专门为这个城市而建造的。这里展示的是它的地质、模型图等，与它在绘画、绘图、照片等方面的表现和美有关。在此背景下，人们逐渐准备了一份关于爱丁堡的调查，从它的史前、起源，到它的不同阶段，直到今天的摄影细节。

以下楼层分别为苏格兰、英国和英语世界、欧洲或西方、东方和人类而修建。

在这座塔所满足的不同功能中，有趣的是将它作为：①天文台，与对现实的直接感知有关；②暗箱，投影是真实的第一个操作；③通过连续的步骤对城市的知识-行为过程进行空间类比：感知、内省（艺术家的观点）、分析（科学的前景）、综合体和传播。

在这座塔展示的内容中，最具特色的是爱丁堡地区的正视图。在某些时候，作者指出了对具体事物的一般看法的重要性，具体事物在他所谓的概要观点中是广受欢迎的综合知识的载体。

梦寐以求的知识综合体（思想家通常在抽象概念中寻求如此之多）是否在我们对具体世界的调查中直接显现在我们周围呢？如果古老的知识大师亚里士多德推动"天象观"的论述被证实确实存在而不仅仅是隐喻呢？当然，"一般的观点"会被一般的观点很好地帮助。

将该塔进一步解读为一个"知识-行动"空间类比的过程是可能实现的。在这个过程的最后一个序列，我们已经将"感知-内省-分析-综合体-传播"描述为街道出口的象征，通过已获得的知识推动现实生活的活动，所以这个序列不是线性的，而是环形的，或者说是螺旋的，这个序列的标志反复地在盖迪斯的方案中所体现。

被分析过的例子有望帮助阐明盖迪斯所谓的视觉思维的意义。更深层次的被盖迪斯所注意的例子也可以被添加进来。举几个例子：盖迪斯在伦敦人口统计图的基础上提出了城市群的概念，这张图纸是Ch. 布斯在盖迪斯对伦敦分析时所绘制的，其中有"对街道和地区的观察研究，对居民的状况和生活的研究"，许多被收集或制作的图像在他所组织的公民展览会上被有目的地展出。关于视觉作为思想和理论的一种感知、分析、推理和交流手段的大量建议，尽管这些建议有其多样性甚至差异性，但在思维方式上显示出一致性，其原则如下。

①以现实为出发点，分析对象，转化行为。

②一般观点的重要性，就像它们的范围、空间或时间以及它们的方法一样，是多学科的。

③教育的重要性在于将获得的知识传递给整个社会，是改善现实的主要杠杆。

对"知识-行为"理解的关注将作为这一切的基础，其以视觉图像为关键工具。

盖迪斯所了解的现实无疑与今天不同，但继续思考城市的性质以及改善城市的机制和工具的必要性是相同的。在我们看来，这就是他的非常重要的遗产。

VOCABULARY

- section ['sekʃən] n. 剖面；截面；部分；部门；地区；章节
- settlement ['setəlmənt] n. 解决；处理；结算；沉降
- depict [dɪ'pɪkt] vt. 描述；描画
- occupation [ˌɒkjə'peɪʃən] n. 职业；占有；消遣；占有期
- stage [steɪdʒ] n. 阶段；舞台；戏剧；驿站

- subsistence [səbˈsɪstəns] n. 生活,生计;生存;存在
- characterized [ˈkærəktəˌraɪzd] adj. 以……为特点的
- evolutive [ˈevəˌluːtɪv] adj. 发展的;进化的
- fertile [ˈfɜːtaɪl] adj. 丰富的;富饶的
- symbolic [sɪmˈbɑlɪk] adj. 象征的;符号的;使用符号的
- standpoint [ˈstændpɔɪnt] n. 立场;观点;立足点
- diagram [ˈdaɪəɡræm] n. 图解;图表
- cloister [ˈklɔɪstər] n. 回廊;修道院;修道院生活;隐居地
- formula [ˈfɔːmjələ] n. 公式,准则;配方;婴儿食品
- reflection [rɪˈflekʃən] n. 反射;沉思
- generalized [ˈdʒenərəlaɪzd] adj. 广义的,普遍的;无显著特点的
- aesthetics [esˈθetɪks] n. 美学
- chord [kɔːd] n. 弦;和弦;香水的基调
- abstract [ˈæbstrækt] n. 摘要;抽象;抽象的概念
- analytic [ˌænəˈlɪtɪk] adj. 分析的;解析的;善于分析的
- literal [ˈlɪtərəl] adj. 文字的;逐字的;无夸张的
- overlook [ˈəʊvəlʊk] n. 忽视;眺望;观察位置   vt. 忽视;宽恕,不计较
- civism [ˈsɪvɪzəm] n. 公德心;公民精神
- itinerary [aɪˈtɪnərəri] n. 旅程,路线;旅行日程
- projection [prəˈdʒekʃən] n. 建筑物的视图(如正视图);投射;规划;突出;发射
- prospect [ˈprɒspekt] n. 前途;预期;景色;可能   v. 勘探
- phase [feɪz] n. 时期
- concrete [ˈkɒŋkriːt] n. 混凝土;具体物;凝结物   adj. 确定的,确实的
- synoptic [sɪˈnɒptɪk] adj. 天气的;概要的

- spiral ['spaɪərəl] *n.* 螺旋；旋涡；螺旋形之物　*adj.* 螺旋形的
- scheme [skiːm] *n.* 计划；组合；体制；诡计　*v.* 密谋，策划
- agglomeration [əˌglɒməˈreɪʃən] *n.* 城市群；聚落；凝聚；结块；附聚
- mechanism [ˈmekənɪzəm] *n.* 机制；机构

EXERCISES

[Choose the right answers]

1. According to the the article, what's the wrong discription of the Valley Section? _____.

    A. Geddes' most known image

    B. The valley model is a continuous plane

    C. Describe your career in terms of tools

    D. Types of settlement represented by their outlines

2. The features of the image do not include _____.

    A. descriptive and symbolic elements

    B. only involving a single complex aspect

    C. the concept of time evolution

    D. complex information

3. Which is the wrong answer about Geddes calls "the city complied"? _____.

    A. Urban activities come from the interaction of people, work and places

    B. On the last level would find place the city of daily action

    C. The formula Town-City represents it

    D. The Cloister is a new synthesis

4. As an image in itself, which of the following features is not

included? _____.

A. Its graphic content is completely textual

B. It shows relations at different levels

C. It expresses an idea of evolution

D. It goes through a phase of discontinuity

5. What's the wrong description of The Outlook Tower? _____.

A. It overlooks the city

B. The tower was built in 1892

C. It's provided with a camera obscura

D. It's described as a civic observatory and laboratory

6. Among the company's functions provides fulfilled by the Tower, which of the following is not? _____.

A. Observatory   B. Camera obscura

C. Spatial analogy   D. Living space

7. Which principles of Proposals concerning the visual is not contained? _____.

A. It is single discipline

B. Object of analysis and of transforming action

C. The general views are important

D. The importance of education is the transmission of knowledge

[Speak your mind]

1. Can you describe The Valley Section with your own understanding? If you can, do it.

2. What are the characteristics of the images in the Valley Section?

3. Although the reality that Geddes knew is undoubtedly different from that of today, but what did you learn from Geddes? Say it out.

Unit 9　Melbourne's Knowledge-based Urban Development

This chapter was published in its original form as:

Yigitcanlar, Tan and O'Connor, Kevin and Westerman, Cara (2008) The making of knowledge cities: Melbourne's knowledge-based urban development experience. Cities 25(2): 63-72.

TEXT

During the 20th century Melbourne was shaped mainly by manufacturing activities. In the new millennium urban processes are now being shaped by the rise of 21st century occupations, which include business analysts, computing professionals, legal professionals, finance managers, media producers, ICT managers, and policy and planning managers. As a result of the spatial urban change in the city, these jobs are concentrated in Melbourne's core. Melbourne City administration is well aware of these urban processes and municipal strategies are already developed and applied for the KBUD of Melbourne. For example, Melbourne City Plan 2010 builds on a vision for the city by focusing on nine key directions mostly concentrating on the liveability and prosperity of the city. The objectives of 2010 Melbourne City Plan reveal some hints about how the city's future is planned as a KC:

①Develop Melbourne as a gateway for biotechnology in Australia and the Asia-Pacific region;

②Redress the skill shortage in the ICT sector and build Melbourne's reputation as the ICT capital of Australia;

③Attract key strategic knowledge industry businesses to move to Melbourne and support and facilitate innovative start-up businesses;

④Promote growth in Melbourne's tertiary education services;

⑤Develop and promote Melbourne as a place that understands, respects and operates successfully with other business cultures;

⑥Develop and promote Melbourne's diverse and highly skilled workforce regionally, nationally and globally to attract global projects;

⑦Enhance and promote Melbourne's liveability and lifestyle options.

The metropolitan strategy plan for Melbourne, much like the City Plan, provides vision for a strong and innovative economy based on the view that all sectors of the economy are critical to economic prosperity. Economic and knowledge clusters play a critical role in the KBUD success of Melbourne. Melbourne 2030 reads that "Opportunities will be protected for internationally competitive industry clusters seeking large landholdings, and for major logistics industries that need ready access to road and rail networks, airports and seaports". This plan also expands logistics and communications infrastructure, including broadband telecommunications services, to underpin development of the innovation economy which is vital to Melbourne's success. In Central Melbourne, the Central Activities District and Docklands are planned to remain a key location for high-order commercial development and the retail and entertainment core of the metropolitan area. Continued residential development in Central Melbourne would take advantage of this area's unmatched accessibility to jobs, facilities, recreational and cultural opportunities, adding to the after-hours vibrancy of the inner areas.

The traces of Melbourne's KBUD are not only evident in the City and Metropolitan Plans. The policies of designing Melbourne as a prosperous global city and a KC date back to late 1970s. Melbourne and Metropolitan Board of Works report on alternative strategies for metropolitan Melbourne, indicating a shift in the denser redevelopment of inner Melbourne which may require a substantial change in housing preferences and lifestyles. This change was part of the new urban containment policy to improve the quality of life and diverse cultural texture and lifestyle options within the city, and address problems of housing affordability. State Government's 1984 economic strategy "Victoria: The Next Steps" identified nine areas which Melbourne has competitive strengths, including Melbourne's world class scientific research institutions. State Government's "1986 Technology Statement" acknowledged the strength of Melbourne's research base and pointed out the urgent necessity of a shift from resource-intensive to knowledge-intensive industries. Beginning from late 1980s Victorian Government started to invest and develop knowledge precincts in the metropolitan Melbourne region. Social Justice Coalition's report on Melbourne's Docklands revealed Melbourne's vision for knowledge precincts and the development of these precincts were seen to provide an effective solution to economic problems. This report examined some of the lessons from overseas experience and discussed the applicability of these models for Melbourne.

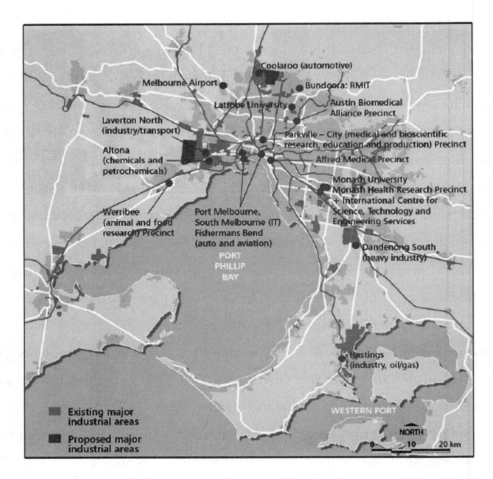

Knowledge clusters of Melbourne metropolitan area (Victorian Government 2002a: 87) This report examined some of the lessons from overseas experience and discussed the applicability of these models for Melbourne.

Similarly Department of Planning and Development saw the prosperity increasingly depending on the ability of Melbourne to compete in the world economy. Melbourne metropolitan strategy acknowledged that the performance of Victoria was depending, to a large extent, on Melbourne's global economic competitiveness and also its ability to operate efficiently as an urban system focused on knowledge creation. At this time, effort was put in to the identification of "knowledge precincts", areas surrounding the main university campuses and had special local land use regulations in favour of high-tech industries, with links to a nearby university. Knowledge precincts have an important role to play in Melbourne's overall R&D infrastructure. They provide opportunities for linkages, technology diffusion and cross fertilisation between high-tech businesses, academia and public sector R&D facilities. Successful precincts act as catalysts for attracting new talent and investment, building up critical mass in particular areas of research and commercialisation. Some of the successful knowledge precincts of Melbourne include Monash, La Trobe, Werribbee, and Port Melbourne/South Melbourne.

To boost sustainable business and trade in Melbourne, Federal, State and City Governments have a number of business development and support funds and programs (e.g. Victorian Biotechnology Strategic Development Plan 2007) available for small, medium size and international companies. Melbourne has a large concentrations of advanced industrial and scientific research in the Asia-Pacific region. The depth of research available is evolving into clusters of cutting-edge expertise not only in academia, but in sectors as diverse as nanotechnology, biotechnology, automotive, aeronautics, financial services and design. Location and employment levels of some of these clusters are given in the picture below.

Melbourne ranks in various listings among the world's most liveable cities (i.e. Economist Intelligence Unit, 2005). Quality of life and place of

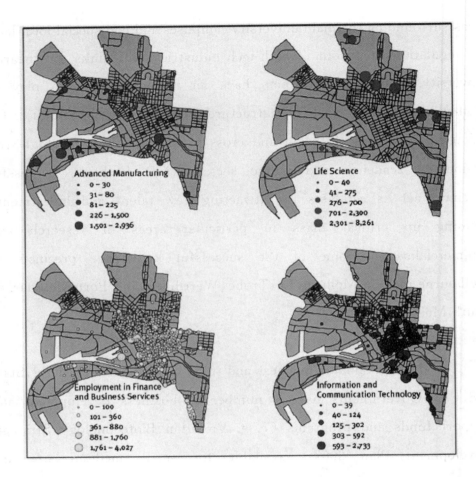

Knowledge clusters of Melbourne City (Melbourne City, 2006: 7)

Melbourne plays a significant role in this recognition, which is achieved through strategising and carefully planning of its urban development and socio-cultural atmosphere. This accomplishment is not only limited to bringing all business, education, R&D clusters together, but also other clusters such as tourism, sports, art and culture have immense contributions to its transition into a KC. On top of these cluster developments alternative inner city lifestyle options such as "Docklands" and "post code 3000" also contributed greatly to the reputation of Melbourne in its long journey to become a globally admired KC.

TRANSLATION

墨尔本的知识城市发展

二十世纪的墨尔本的城市发展主要受制造业活动的影响。在新的一千年,城市进程被21世纪的新兴职业所影响,包括商业分析师、计算专业人士、法律专业人士、财务经理、媒体制作人、信息和通信技术人员以及政策和规划管理人员。作为城市空间演变的结果,这些工作地点集中在墨尔本的市中心。墨尔本市管理部门非常重视这些城市发展进程,制定并应用了多种行政策略用于墨尔本的知识城市发展。例如,2010年墨尔本城市规划通过聚焦九个关键方向来建立一个城市愿景,主要关注宜居性和城市活力。2010年墨尔本城市规划的目标揭示了作为知识城市的未来方向:

①将墨尔本发展成为澳大利亚和亚太地区生物技术的门户;

②弥补信息和通信技术行业的短板,并打造墨尔本作为澳大利亚信息和通信技术都城的声誉;

③吸引关键战略知识行业迁往墨尔本,支持和促进创新型初创企业;

④促进墨尔本高等教育服务的增大;

⑤发展和推动墨尔本成为一个理解、尊重并与其他商业文化成功运作的地方;

⑥在区域、国家和全球范围内开发和推广墨尔本多元化和高技能的劳动力,以吸引全球项目;

⑦加强和促进墨尔本宜居和生活方式的选择。

与城市规划非常相似,墨尔本的大都市区战略规划基于所有经济部门都对经济繁荣至关重要的观点,为强大和创新的经济提供了未来发展愿景。经济和知识集群在墨尔本知识经济的成功发展中发挥了关键作用。

《墨尔本2030》提出"保护国际性竞争性产业集群寻求大量用地的机会，同时保护主要物流产业与公路、铁路网络、机场和海港建立便捷联系的机会"。该规划还扩展了包括宽带电信服务在内的物流和通信基础设施服务，以支持创新经济的发展，这对墨尔本的成功至关重要。在墨尔本市中心，中央活动区和码头区计划仍然是发展高端商业开发，以及作为大都市区零售和娱乐核心的重要地点。墨尔本市中心的持续住宅开发将充分利用该地区极好的就业、公共服务设施、生活娱乐和文化机会，增加内陆地区的业余时间的活力。

墨尔本知识城市发展的痕迹不仅体现在城市规划和大都市区规划中。将墨尔本设计为繁荣的全球城市和知识城市的政策可以追溯到20世纪70年代末。墨尔本和大都会工作委员会曾经报告了墨尔本都市区的一个替代性开发战略，提出在墨尔本内城进行密集的重建，可能需要在住房偏好和生活方式上做出重大改变。这一转变是新的城市容量政策的一部分，用于提升城市生活质量、丰富城市肌理和生活方式选择，并且解决城市居民住房问题。州政府1984年的经济战略《维多利亚：下一步》规划中确定了墨尔本具有竞争优势的九个领域，包括墨尔本的世界级科研机构等。州政府的《1986年技术声明》中强调了墨尔本研究基地的实力，并指出了从资源密集型产业向知识密集型产业转变的迫切必要性。从20世纪80年代末开始，维多利亚州政府开始在墨尔本大都市区投资和开发。社会公益组织关于墨尔本码头区的报告显示，墨尔本对知识区的规划和这些区域的发展是解决经济问题的有效手段。该报告还研究了一些海外案例，并讨论了这些模型在墨尔本的适用性。

同样，城市规划和发展部门意识到成功越来越取决于墨尔本在世界经济中的竞争能力。墨尔本大都市区战略承认维多利亚州的表现在很大程度上取决于墨尔本的全球经济竞争力，同时还有它作为一个城市系统有效控制知识创新的能力。此时，城市做出努力来识别"知识区"，即那些环绕在主要大学校园周边，以及与附近的大学联系紧密，制定了有利于高科技产业的地方土地使用规定的区域。知识区在墨尔本的整体研发设施布局中发挥着重要作用。知识区为高科技企业、学术界和公共部门与研发机构之间的联系，为技术传播和交叉融合提供了平台。成功的知识区可以吸引新的人才和投资，在特定的研究和商业化领域建立起重要区域。墨尔本的一些成功的知识区包括蒙纳士，拉筹伯，华勒比和墨尔本港/南墨尔本。

为了促进墨尔本商业和贸易的可持续发展，联邦、州和市政府拥有一些业务开发和支持基金和计划（例如 2007 年维多利亚州生物技术战略发展计划），可用于中小型和国际公司。墨尔本是亚太地区很大的先进工业和科学研究集中地。墨尔本的研究领域已从学术界扩展到纳米技术、生物技术、汽车、航空、金融服务和设计等多个领域，逐渐发展成为一系列尖端专业知识集群。下图（见前文图）给出了其中一些集群的位置和就业水平。

墨尔本在许多世界上最适宜居住的城市排名中都榜上有名（例如，经济学人智库，2005）。墨尔本的生活质量和城市所在地在这一认可中起着重要作用，墨尔本获得这一称号是通过规划制定合理的城市发展战略和社会文化氛围的精心营造来实现的。它不仅限于将所有商业、教育、研发集群整合在一起，而且也囊括了包括旅游、体育、艺术和文化在内的其他集群，为其向知识城市过渡做出了巨大贡献。除了这些集群发展之外，多样化的城市生活方式选择，如"码头区"和"邮编 3000"区，也在墨尔本成为全球知名知识城市的漫长旅程中，为墨尔本的城市影响力的形成做出了巨大贡献。

VOCABULARY

- clarity [ˈklærəti] *n.* 清晰；清楚；明确
- spatial [ˈspeɪʃəl] *adj.* 空间的；存在于空间的；受空间条件限制的
- cluster [ˈklʌstər] *n.* 群集；簇；丛
- landholding [ˈlændˌhəʊldɪŋ] *adj.* 土地所有的；占用土地的 *n.* 拥有土地；占有土地
- metropolitan area 大都市区
- containment [kənˈteɪnmənt] *n.* 容量；包含；牵制
- precinct [ˈpriːsɪŋkt] *n.* 选区；管理区；管辖区
- catalyst [ˈkætəlɪst] *n.* 催化剂；刺激因素
- commercialization [kəˌmɜːʃəlaɪˈzeɪʃən] *n.* 商品化，商业化
- cutting-edge [ˌkʌtɪŋˈedʒ] *adj.* 先进的，尖端的 *n.* 尖端；前沿；刃口
- expertise [ˌekspɜːˈtiːz] *n.* 专门知识；专门技术；专家的意见

EXERCISES

[Choose the right answers]

1. According to the first paragraph, in the new millennium Melbourne was shaped mainly influenced by _____.

    A. manufacturing activities

    B. various kinds of fresh jobs

    C. policy and planning managers

    D. secondary industry

2. Melbourne 2030 illuminates that _____.

    A. clusters seeking large landholdings is difficult in Melbourne

B. internationally competitive industry clusters and logistics industries will be supported

C. logistics industries cause serious pollution problems in the city

D. internationally competitive industry clusters don't need ready access to road and rail networks, airports and seaports

3. The influence of Melbourne's KBUD are related to different kinds of Plans except _____.

A. City Plans

B. Metropolitan Plans

C. alternative strategies for metropolitan Melbourne

D. denser redevelopment

4. _____ provide chances for linkages, technology diffusion and cross fertilisation between high-tech businesses, academia and public sector R&D facilities.

A. Melbourne's overall R&D infrastructure

B. Knowledge precincts

C. Metropolitan strategy

D. Successful precincts

5. Melbourne is regarded as "the most liveable city" because of _____.

A. its high quality of life and good environmental resources

B. the carefully plans of urban development

C. its social-cultural atmosphere

D. All the above

[Speak your mind]

1. What impact did the appear of emerging professions in the 21st

century have on Melbourne's urban space?

2. What played a key role in the KBUD success of Melbourne?

3. What was the main content of alternative strategies for metropolitan Melbourne proposed by Melbourne and Metropolitan Board in the late 1970s?

4. In order to realize the transition to KC, in addition to integrating all business, education and R&D clusters, what has Melbourne done?

Unit 10   Greenbelt of Tokyo

This chapter was published in its original form as:

M. Yokohari et al. (2000), Beyond Greenbelts and Zoning: A New Planning Concept for the Environment of Asian Mega-Cities, Landscape and Urban Planning ,Urban Ecology,159-171.

TEXT

When observing contemporary urban landscapes in Asian mega-cities,one may hardly realize that there have been attempted to apply western planning methods on land use to keep explosive urban expansion under control. Chaotic landscapes identified in the fringe of mega-cities are a clear examples that document the absence of effective controls. However, Asian mega-cities did, and still do, have physical urban plans including land use and zoning plans. Greenbelt is commonly applied concepts to Asian mega-cities.

Tokyo installed a comprehensive parks and open space master plan in 1939. The plan included parks and open space in various scales in Greater Tokyo area of 9600 km^2; from urban parks, cemeteries and allotment gardens in the central district to scenic beauty areas and national parks in remote mountains. The plan is regarded as the most ambitious plan in the history of parks and open space plans in Japan.

The plan included a greenbelt on the boundary of the Ward Area of Tokyo. The Amsterdam Declaration in 1924 by the International Federation of Housing and Planning (IFHP), which identified the need for establishing greenbelts when planning for urban expansion, was the theoretical basis of the installation. The greenbelt, total 136 km^2, consists of farmland and coppice woodland, was planned on a 15 km radius to restrict disordered expansion of densely inhabited urban areas. The belt was associated with radial green corridors planned along river ravines flowing into downtown. Recreational paths such as pedestrian and horse riding trails were planned in these corridors.

Succeeding the 1939 plan, a new open space plan for Tokyo was decided in 1943 to meet the needs of air defense during the World War II. The concept of the plan was to create open areas to stop the spread of fire caused by bombing and provide refuge and escape routes. The focus was to create green corridors. In addition to the greenbelt, an inner circular corridor was planned on a 10 km radius to surround urbanized area at the time by connecting major urban parks planned in the 1939 plan. Radial corridors along river ravines connected outer and inner radial corridors. The double circular and radial fluvial corridors in Tokyo reached 123.5 km.

The air defense open space plan was terminated and succeeded by the post-war rehabilitation open space plan of 1947. In this plan, the focus was again given to the creation of circular and radial corridors. The double-ring circular green corridors, including a greenbelt and a network of radial green corridors along trunk roads, rivers and railroads were planned to connect urban parks.

If the plan was fully implemented, central Tokyo might have been the richest green cities in the world with over 200 km² of green spaces in the central district. However, as the urban landscape of Tokyo today clearly represents, the plan was poorly implemented. Only a few fluvial corridors were realized, while the circular green corridor gradually decreased and completely abolished in 1969. Today, only 4%, 24 km², of the Ward Area is ceded as parks and open space.

TRANSLATION

东京的绿带

在观察亚洲大城市的当代城市景观时,人们可能很难意识到这一点,那就是过去曾试图用西方的土地利用规划方法来控制爆炸性的城市扩张。大城市边缘景观混乱是一个表明该城市缺乏有效控制的清晰的例子。然而,亚洲大城市一直以来都有实体的城市规划,包括土地使用和分区规划。绿带是亚洲大城市常用的概念。

1939年,东京建立了综合公园和开敞空间的总体规划。该规划包括东京都市圈9600 km²的各种规模的公园和开敞空间;还包括从市中心的城市公园、墓地、园林到风景名胜区和偏远山区的国家公园。该规划被认为是在日本有史以来公园与开敞空间规划中的最富有雄心壮志的规划。

规划包括在东京特别区边界的绿带。1924年国际住房与规划联合会(IFHP)的《阿姆斯特丹宣言》指出,在规划城市扩张时建立绿带的必要性是该设施的理论基础。绿带总面积为136 km^2,由农田和矮林组成,规划半径为15 km,以限制人口密集的城区无序扩张。这条绿带与沿着河流沟壑流入市中心的放射状绿廊相关联。在这些廊道中还规划了诸如步行道和骑马道等休闲路径。

继1939年规划之后,为了满足第二次世界大战期间的防空需求,东京在1943年制定了一个新的开敞空间规划。该规划的概念是建立开敞区域,来阻止爆炸引起的火势蔓延,并提供避难和逃生路线。重点是创建绿色走廊。除了绿带以外,还规划了一条半径为10 km的内环线走廊,通过连接1939年规划中的主要城市公园,环绕当时的城市化地区。沿着河流沟壑的放射状廊道连接内外放射状廊道。东京的双环形和放射状河流走廊长达123.5 km。

防空开敞空间规划被终止,由1947年的战后复兴开敞空间规划所继承。这一规划中的重点再次放到建立环形放射状走廊。双环形绿色走廊在规划中让其连接城市公园,双环形绿色走廊包括一条绿带和沿主干道、河流与铁路的辐射状绿色走廊网络。

如果该规划得以全面实施,东京市中心可能成为世界上最富有的绿色城市,中心区的绿地面积超过200 km^2。然而,正如今天东京的城市景观所展现的那样,这个规划实施得很糟糕。只有少数河流走廊得以实现,而环状绿色走廊逐渐减少并于1969年完全被废除。今天,只有4%,也就是24 km^{2}的特别区被划为公园和开敞空间。

VOCABULARY

- greenbelt [ˈgriːnˌbelt] n. 绿带;(城市周围的)绿色地带
- observe [əbˈzɜːv] vt. 观察;遵守;说;注意到;评论
- contemporary [kənˈtempərəri] adj. 当代的;同时代的;属于同一时期的
- mega-city 巨型城市
- explosive [ɪkˈspləʊsɪv] adj. 爆炸的;爆炸性的;爆发性的
- expansion [ɪkˈspænʃn] n. 膨胀;阐述;扩张物
- chaotic [keɪˈɒtɪk] adj. 混沌的;混乱的,无秩序的
- identified [aɪˈdentɪfaɪd] adj. 被识别的;经鉴定的
- absence [ˈæbsəns] n. 没有;缺乏;缺席;不注意
- zoning [ˈzəʊnɪŋ] n. (美)分区制;都市的区域划分
- install [ɪnˈstɔːl] vt. 安装;任命;安顿
- master [ˈmɑːstər] adj. 主要的;熟练的;主人的 n. 主人;大师,名家
- scale [skeɪl] n. 规模;比例;鳞;刻度;天平;数值范围
- cemetery [ˈsemətri] n. 墓地;公墓
- allotment [əˈlɒtmənt] n. 分配;分配物;养家费;命运
- district [ˈdɪstrɪkt] n. 区域;地方;行政区
- scenic [ˈsiːnɪk] adj. 风景优美的;舞台的;戏剧的
- ambitious [æmˈbɪʃəs] adj. 野心勃勃的;有雄心的;热望的;炫耀的
- boundary [ˈbaʊndəri] n. 边界;范围;分界线
- theoretical [θɪəˈretɪkəl] adj. 理论的;理论上的;假设的;推理的
- installation [ˌɪnstəˈleɪʃn] n. 安装,装置;就职
- coppice [ˈkɒpɪs] n. 矮林;小灌木林 vt. 修剪(树木或灌木)使成

为萌生林
- woodland ['wʊdlənd] *n.* 林地；森林
- radius ['reɪdiəs] *n.* 半径，半径范围；辐射光线；有效航程；桡骨
- associate [ə'səʊsieɪt] *v.* 联想；(使)发生联系；(使)联合；结交
- radial ['reɪdiəl] *adj.* 半径的；放射状的；光线的；光线状的
- corridor ['kɒrɪdɔːr] *n.* 走廊，通道；走廊
- ravine [rə'viːn] *n.* 沟壑，山涧；峡谷
- downtown ['daʊn'taʊn] *n.* 城市中心区
- pedestrian [pə'destriən] *n.* 行人；步行者
- defense [dɪ'fens] *n.* 防卫，防护；防御措施；防守
- bombing ['bɒmɪŋ] *n.* 轰炸，投弹
- refuge ['refjuːdʒ] *n.* 避难；避难所；庇护
- fluvial ['fluːviəl] *adj.* 河流的；生在河中的；河流冲刷形成的
- terminate ['tɜːmɪneɪt] *v.* 结束，终止；结果
- rehabilitation [ˌrihəˌbɪlə'teɪʃən] *n.* 修复；复兴；复职；恢复名誉
- implement ['ɪmplɪmənt] *v.* 实施，执行

EXERCISES

[Choose the right answers]

1. What is the clearest example that indicates the absence of effective controls? _____.

A. There have been attempted to apply western planning methods on land use

B. Asian mega-cities hardly realize the importance to keep explosive urban expansion under control

C. Chaotic landscapes identified in the fringe of mega-cities

D. Asian mega-cities have physical urban plans including land use and zoning plans

2. Tokyo comprehensive parks and open space master plan include except _____.

A. parks and open space in various scales

B. cemeteries

C. allotment gardens in the central district

D. county parks

3. According to the description of Tokyo comprehensive plan, we can know the information below except that _____.

A. the greenbelt covers total 136 km²

B. the greenbelt was planned on a 15 km diameter

C. the greenbelt is consists of farmland and coppice woodland

D. the greenbelt aims to restrict disordered expansion of densely inhabited urban areas

4. The focus of the new open space plan for Tokyo was to _____.

A. meet the needs of air defense

B. create open areas to stop the spread of fire

C. provide refuge and escape routes

D. create green corridors

5. The double-ring circular green corridors were planned to _____.

A. ensure the plan will be fully implemented

B. connect major urban parks planned in the 1939 plan

C. connect urban parks

D. create circular and radial corridors

[Speak your mind]

1. What is the common applied concepts in Asian mega-cities to keep explosive urban expansion under control?

2. What are the characteristics of the greenbelt on the boundary of the Ward Area of Tokyo?

3. What are the concept and focus of Tokyo's new open space plan in 1943?

4. What do the double-ring circular green corridors include in the post-war rehabilitation open space plan of 1947?

5. Please illustrate the present implement situation of Tokyo's post-war rehabilitation open space plan.

Unit 11　Planning Policy: The London Plan

This chapter was published in its original form as:

T. Beatley(2012), Green Cities of Europe: Global Lessons on Green Urbanism, Island Press, 186-188.

TEXT

In the United Kingdom there is a clear and consistent hierarchy of legislation. Overarching policy direction is set by the European Union. National planning policy is high level and must be in conformity with the EU, and regional policy (such as the London Plan) must be in conformity with national policy. At the most local level within London, each of the thirty-three boroughs must have local development plans (called Local Development Frameworks) that must be "in general conformity" with the London Plan.

Creating a strong policy context for growth is central to London's sustainability approach. The Greater London Authority Act (1999) requires the GLA to produce, and keep under review, a spatial development strategy for London, known as the London Plan. It should be an overall strategic plan, setting out an integrated economic, environmental, transport, and social framework for the development of London over the next twenty-five years. The act of Parliament requires that the London Plan take account of three cross-cutting themes: economic development and wealth creation, social development (including crime prevention), and improvement of the environment. The preparation of the plan requires an Integrated Impact Assessment, which includes the legal requirements to carry out a Sustainability Appraisal (this includes a Strategic Environmental Assessment and a Habitats Regulation Assessment), and to ensure that health, equality, and community safety are properly handled. Prior to a plan being adopted, it is subjected to an examination in public, led by an independent panel, which scrutinizes the document and reviews comments from interested citizens. This process is intended to enable public participation in the plan's preparation, and reflects the principles in the EU Aarhus Convention on access to information, public participation, and access to justice in environmental matters. The panel recommends changes that the mayor can consider when finalizing the plan, which is then submitted to the Government Office for London, where ministers decide whether to instruct any further changes prior to the plan being formally adopted.

The first London Plan was published in 2004 and revised in 2008. The current London Plan, which sets out the policy to 2031, was adopted in 2011. It is a comprehensive suite of interrelated policies to support Mayor Boris Johnson's vision for London: "London should excel among global cities—expanding opportunities for all its people and enterprises, achieving the highest environmental standards and quality of life and leading the world in its approach to tackling the urban challenges of the

21st century, particularly that of climate change."

The plan sets out to ensure that development is sustainable and that climate change is tackled. It seeks to protect London's natural resources, environmental and cultural assets, the health of its people, and to adapt to and mitigate the effect of climate change. These are covered in six key objectives, ensuring that London is:

* a city that meets the challenges of economic and population growth in ways that ensure a sustainable and improving quality of life for all Londoners, and helps tackle the huge issue of inequality among Londoners, including inequality in health;

* an internationally competitive and successful city, with a strong and diverse economy and an entrepreneurial spirit that benefits all Londoners and all parts of London; a city that is at the leading edge of innovation and research; and that is comfortable with—and makes the most of—its rich heritage and cultural resources;

* a city of diverse, strong, secure, and accessible neighborhoods to which Londoners feel attached, which provides all its residents, workers, visitors, and students—whatever their origin, background, age, or status—with opportunities to realize and express their potential, and a high-quality environment for individuals to enjoy, to live in together, and to thrive in;

* a city that delights the senses and takes care of its buildings and

streets, having the best of modern architecture while also making the most of London's built heritage, and which makes the most of and extends its wealth of open and green spaces and waterways, realizing its potential for improving Londoners' health, welfare, and development;

* a city that becomes a world leader in improving the environment locally and globally, taking the lead in tackling climate change, reducing pollution, developing a low-carbon economy, and consuming fewer resources and using them more effectively;

* a city where it easy, safe, convenient for everyone to access jobs, opportunities, and facilities with an efficient and effective transport system that actively encourages more walking and cycling, makes better use of the Thames, and supports delivery of all the objectives of this plan.

TRANSLATION

规划政策：伦敦规划

在英国，有一个明确和一致的立法层级。总体政策方向由欧盟确定，国家规划政策是高层级的，必须与欧盟相适应，区域政策（如伦敦规划）必须遵照国家政策。在伦敦最基层，三十三个行政区中的每一个区都必须制定当地的发展规划（称为"地方发展框架"），这些规划必须与伦敦规划"基本一致"。

为增长创造一个强有力的政策背景是伦敦可持续发展策略的核心。《大伦敦政府法案》(1999年)要求大伦敦政府为伦敦制定并不断审查一项被称为"伦敦规划"的空间发展战略。它应该是一项总体战略规划,为未来二十五年伦敦的发展制定一个完整的经济、环境、交通和社会框架。《国会法案》要求伦敦规划考虑三个交叉主题,分别是经济发展和财富创造、社会发展(包括预防犯罪)、环境改善。该规划的编制需要进行综合影响评估,其中包括进行可持续性评估的法律要求(包括战略环境评估和栖息地监管评估),并确保健康、平等和社区安全得到妥善处理。在一项规划被采纳之前,它将由一个独立专家组带领进行公开审查,该专家组审查文件并审查感兴趣的公民的评论。这一进程旨在使公众参与该规划的编制工作,并反映《欧盟奥胡斯公约》在环境问题上关于信息获取、公众参与和通向司法公正的原则。专家组建议,市长在最终敲定该规划时可以考虑进行哪些修改变更,然后提交给伦敦政府办公部门,由部长们决定是否在正式通过该规划之前指示任何进一步的变更。

第一个伦敦规划于 2004 年公布,并于 2008 年修订。现行的伦敦规划制定了一直到 2031 年的政策,并于 2011 年通过。这是一整套相互关联的政策,以支持市长鲍里斯·约翰逊对伦敦的愿景:"伦敦应该在全球城市中脱颖而出——为所有的人和企业扩大机会,实现最高的环境标准和生活质量,引领世界应对 21 世纪的城市挑战,特别是应对气候变化方面的挑战。"

该规划旨在确保发展是可持续的,并解决气候变化问题。它寻求保护伦敦的自然资源、环境和文化资产、人民的健康,并适应和减轻气候变化的影响。这些目标涵盖六个关键目标,以确保伦敦能够实现:

＊一个能够满足经济和人口增长挑战的城市,它能确保所有伦敦人的生活质量持续提高,并有助于解决伦敦人与人之间不平等的巨大问题,包括健康方面的不平等;

＊一个具有国际竞争力和成功的城市,往往拥有强大而多元化的经济

和创业者精神，使所有伦敦人和伦敦各地都受益；一个处于创新和研究前沿的城市；这座城市充分利用了其丰富的文化遗产和文化资源，这让人感到舒适；

＊一个拥有多元化、坚固、安全和通达的社区城市，为所有居民、工人、访客和学生，无论其出身、背景、年龄或地位，提供实现和表达潜力的机会，以及为个人提供一个高质量的环境来享受，并且在这个环境下能共同生活并茁壮成长；

＊一个使感官享受并关注建筑和街道的城市，它拥有最好的现代建筑，同时也充分利用了伦敦的建筑遗产，最大限度地发挥了它丰富的开敞空间、绿色空间和水路的作用，提高伦敦人的身体素质和福利，并且使伦敦人的发展得以提升；

＊一个在当地和全球改善环境方面成为世界领导者的城市，这个城市率先应对气候变化，致力于减少污染、发展低碳经济，争取消耗更少的资源并更有效地利用资源；

＊在这个城市里，每个人都可以轻松、安全、方便地获得工作、机会和设施，拥有高效的交通系统，积极鼓励更多的步行者和骑自行车的人，更好地利用泰晤士河，并支持这项规划的所有目标。

VOCABULARY

- hierarchy [ˈhaɪərɑːki] *n.* 等级制度；统治集团
- overarching [ˌəʊvəˈrɑːtʃɪŋ] *adj.* 首要的；包罗万象的，支配一切的
- conformity [kənˈfɔːməti] *n.* 遵从，遵守；从众；符合，符合度
- borough [ˈbʌrə] *n.* 市镇；大城市中的行政区
- sustainability [səˌsteɪnəˈbɪləti] *n.* 持续性，能维持性，永续性
- spatial [ˈspeɪʃəl] *adj.* 空间的
- strategy [ˈstrætədʒi] *n.* 战略；策略；计谋；行动计划；策划，部署

- overall [ˌəʊvəˈrɔːl] adj. 总的；全面的；包括一切的
- integrated [ˈɪntɪɡreɪtɪd] adj. 综合的，完整统一的
- framework [ˈfreɪmwɜːk] n. 架构，框架；体系，结构
- prevention [prɪˈvenʃən] n. 阻止，妨碍；预防
- community [kəˈmjuːnəti] n. 社区；群体；社团，团体
- independent [ˌɪndɪˈpendənt] adj. 独立的；无党派的；自立的，独立的
- scrutinize [ˈskruːtɪnaɪz] vt. 细看，仔细审查
- public participation 公众参与
- recommend [ˌrekəˈmend] vt. 推荐，介绍；建议
- minister [ˈmɪnɪstər] n. 部长；大臣；牧师；公使；外交使节
- instruct [ɪnˈstrʌkt] vt. 指示，命令，吩咐；委托；作说明；讲，教授
- vision [ˈvɪʒən] n. 幻景；幻象，幻觉；眼光；远见；眼力；视力；视觉
- enterprise [ˈentəpraɪz] n. 组织；公司，企业
- asset [ˈæset] n. 财产；优点，长处；有用的人；资产
- mitigate [ˈmɪtɪɡeɪt] vt. 使缓和；减轻
- tackle [ˈtækəl] vt. 对付，处理；与……交涉；阻截
- inequality [ˌɪnɪˈkwɒləti] n. 不平等，不均等
- competitive [kəmˈpetɪtɪv] adj. 具有竞争力的
- entrepreneurial [ˌɒntrəprəˈnɜːriəl] adj. 企业家的，创业者的；中间商的
- heritage [ˈherɪtɪdʒ] n. 遗产
- resident [ˈrezɪdənt] n. 居民；住户
- welfare [ˈwelfeər] n. 福利救济；社会福利
- globally [ˈɡləʊbəli] adv. 全球地；全局地；世界上
- low-carbon 低碳（的），含碳低的；低碳
- efficient [ɪˈfɪʃənt] adj. 效率高的；有能力的；有效的；生效的

- effective [ɪˈfektɪv] *adj*. 能产生预期结果的；有效的
- delivery [dɪˈlɪvəri] *n*. 运送，递送，投递

EXERCISES

[Choose the right answers]

1. Regional policy must be in conformity with _____.

   A. overarching policy

   B. London Plan

   C. national policy

   D. local development frameworks

2. The London Plan should take account of three cross-cutting themes except _____.

   A. improvement of the environment

   B. social development (including crime prevention)

   C. economic development and wealth creation

   D. an Integrated Impact Assessment

3. _____ is not the principle that reflected by the EU Aarhus Convention.

   A. Access to information

   B. Health and equality

   C. Public participation

   D. Access to justice in environmental matters

4. The current London Plan was adopted in _____.

   A. 2004　　　　B. 2008　　　　C. 2010　　　　D. 2011

5. The underlined phrase "The plan" in the fourth paragraph refers to _____.

A. the first London Plan that published in 2004

B. the current London Plan that sets out the policy to 2031

C. the London Plan that revised in 2008

D. the plan that submitted to the Government Office for London

[Speak your mind]

1. What kind of the frameworks for the development of The Greater London Authority Act require to set?

2. Before a plan is adopted, it will be open for review by an independent panel of experts. What is its purpose?

3. What is central to London's sustainability approach?

4. As a city which has become a world leader, what should London do for local and global environmental improvement?

Unit 12   Midtown Manhattan's Office Projections

This chapter was published in its original form as:

Regional Plan Association (1969), Urban Design Manhattan: A Report of the Second Regional Plan. New York: The Viking Press. 68-73.

TEXT

Midtown Manhattan should be prepared for additional office construction, during the next thirty-five years, equivalent to about 66 Time and Life buildings, eight Rockefeller Centers or eight World Trade Centers. This is about 400,000 office jobs, 80 percent of the increase in office employment projected for the entire Manhattan CBD.

Factors underlying this recommendation are the following.

①Economic projections for the Region indicating a growth of about 1.4 million jobs in office buildings in the 31-county New York Region Study Area. (Regional Plan Association, The Region's Growth, May 1967)

②A finding that, of the total office jobs expected in the Region, a significant share will be head quarters types-related to over-all business polities of large organizations. These have traditionally favored the Manhattan CBD.

③The determination that careful application of the functional, form and amenity principles described in the previous chapter would make possible the accommodation of these 400,000 mere office workers in Midtown, without sacrificing efficiency, convenience or environmental quality.

④The likelihood that this tremendous quantity of offices cannot be properly handled outside the CBD and finally, that the social needs of the Region require keeping a large number of jobs in the older cities. (Regional Plan Association, Jamaica Center, April 1988)

Because one service-type job is usually required to support every five office-type jobs, about 80,000 additional workers must be provided for. An increase can also be expected in visitors for business, shopping and recreation purposes.

PROJECTED OFFICE GROWTH

New office space needed in Mid-town Manhattan to accommodate 400,000 additional office workers equals about 80 million net rentable square feet, symbolized here by 66 Time and Life buildings. The Time and Life building is selected to illustrate the magnitude of projected office expansion in Midtown Manhattan. Built in 1959, it has 48 floors, and employs over 6,000 workers on 1,500,000 square feet of gross rentable office space. This is substantially more than the "average" post-war office building (which has 30 floors and somewhat over 500,000 square feet of rentable space), but is used here, because of distinctive size and coverage (half a large city block) as a symbol of the office space to be added.

Future buildings will vary in size, of course, and are neither expected nor desired to be located in the diagrammatic arrangement on the map at right.

PERMANENT AREAS IN MIDTQINN COMMERCIAL DISTRICT

PLACES WHICH ARE MOST LIKELY TO REMAIN IN PRESENT USE AND PLACES WHICH SHOULD NOT BE CONSIDERED POTENTIAL OFFICE SITES

Given the projected number of new office buildings, where should they be located? Outlined above is the area in Midtown which is presently zoned to accommodate large office buildings. All of this area is in some use now, but some structures are clearly more resistant to change than others, and some should be made more resistant for social and amenity reasons. All colored parcels on the map are occupied by uses which are considered relatively permanent ("hard"). The first four categories shown in the legend are "hard" by fairly objective criteria: all post-1948 construction is expected to endure until the end of the century; major pre-war buildings, such as those of Rockefeller Center, the Chrysler Building and similar structures are also included in this category. Also considered objectively "hard" are important shopping districts, such as Fifth Avenue and Herald Square, as well as major institutions and officially-designated landmarks. The last two categories in the legend are "hard" for relatively subjective social and design reasons. These comprise theaters, clubs, hotels (mostly in the commercial entertainment district) as well as potential landmarks and intimate-scale buildings with specialty shops and restaurants which seem worth retaining despite their susceptibility to displacement with offices or other uses with a greater earning power.

Unit 12 Midtown Manhattan's Office Projections 133

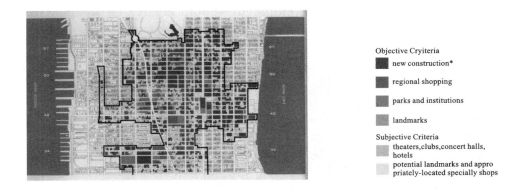

Objective Cryiteria
- new construction*
- regional shopping
- parks and institutions
- landmarks

Subjective Criteria
- theaters, clubs, concert halls, hotels
- potential landmarks and appropriately-located specially shops

"HARD" AND "SOFT" AREAS IN MIDTOWN COMMERCIAL DISTRICT

PERMANENT PLACES AS CONTRASTED WITH IMPERMANENT AREAS SUBJECT TO DEVELOPMENT WITH OFFICE BUILDINGS

The parcels in areas which are not colored in the map on the opposite page contain uses considered impermanent, or "soft". These places are generally most available for development; on the map above, they are shown in blue. Most of Midtown is comprised of uses which seem to be or ought to be relatively permanent, as evidenced by the amount of red above. But there is a great deal of land which is subject to change. Theoretically, there is sufficient land capacity within the area zoned for offices to accommodate most of the projected growth. In practice, however, all the "soft" parcels shown above cannot be assembled for office buildings given present real estate practices, nor should they. Given the design goal of maintaining contrast between "high" and "low", certain areas, notably the theater district, should remain "low" despite the numerous "soft" parcels they contain. Moreover, as the map on the following page shows, much of the "soft" land is not in close proximity to subway stations. Priority in development should clearly go to the "soft" sites with the greatest access potential.

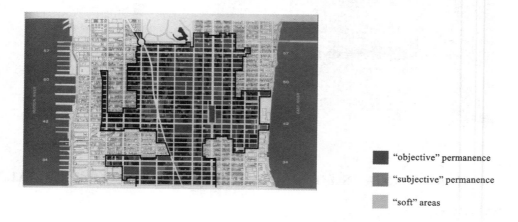

"objective" permanence
"subjective" permanence
"soft" areas

ACCESSIBILITY AS A FORM DETERMINANT
"SOFT" AREAS WITHIN A SHORT WALK OF SUBWAY STATIONS

Large buildings in which several thousand people work should be located within easy walking distance of rail and subway stations. The map above overlays the areas within a three-minute walk of subway stations, shown on page 69, on the map of "soft" or developable parcels shown on the previous page. The resulting areas of overlap are prime sites for the growing of "access trees" within the district presently zoned for office development. These are also priority sites for the implementation of development controls, described in Chapter 5. While adequate for the short to medium range, these sites, however, would not be able to accommodate all of the projected development. For the longer range, sites outside the presently-zoned Midtown office district will have to be sought.

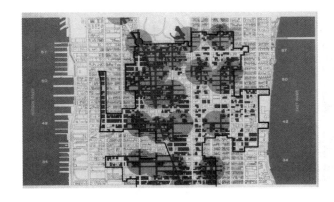

■ "soft" areas

■ areas within 700 feet of subway stations

LAND ASSEMBLY AS A FORM DETERMINANT
PUBLICLY-KNOWN LAND OWNERSHIP

Land ownership within the presently-zoned Midtown commercial district is highly fractionalized. There are very few parcels of one-half acre or more of "soft" land within it that are publicly known to be in single ownership. The greatest concentration of large, developable parcels in single ownership is on the West Side toward the Hudson River, as the above map shows. On this basis, an expansion of the office district toward the Hudson River in the long range would seem to merit investigation. A future westward expansion of the Midtown CBD is, moreover, logically on transportation grounds. Any substantial increment to total CBD employment will require new rapid transit capacity beyond present programs. In view of the geography of Manhattan, the shortest path in and out of it is in an east-west direction. A new high-speed east-west line in Midtown would not only facilitate sorely inadequate crosstown movement and stimulate the commercial development of high-amenity waterfront sites, but also focus residential development into nearby, but underutilized areas to the east and west of Manhattan.

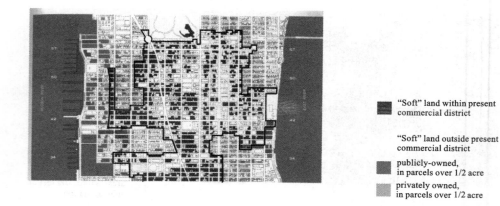

"Soft" land within present commercial district

"Soft" land outside present commercial district

publicly-owned, in parcels over 1/2 acre

privately owned, in parcels over 1/2 acre

The Conceptual Diagram

Midtown's conceptual organization is developed here from the functional and visual form principles discussed in Chapter 3, and in reference to the land use and transportation analysis presented on the previous pages. Clusters of high office towers centered on transportation access points are separated by "lows" in which restaurants, specialty shops, theaters, clubs and department stores are located.

Existing office clusters and possible locations for new clusters are shown in the diagram. Existing office "highs" can be identified: Grand Central Station, the Park Avenue strip, Rockefeller Center and Penn Station. Possible new office clusters would include one on the West Side, expansion and consolidation of Rockefeller Center westward to about Eighth Avenue and an extension of the Penn Station area.

High-density residential clusters are proposed at the four corners of Mid-town. There is a tendency toward this pattern now.

Public open space envelopes the superliner terminal on the Hudson, and extends over and beyond the peripheral highways on both rivers.

A conceptual diagram is but one step toward superior urban design. It provides a formal framework for the many small decisions which may then become a meaningful visual whole.

Pure compositional organization of urban forms and spaces is not the objective of this concept. Its primary objective is to achieve visual clarity, order and maximum amenity.

"Laws" should be retained-on Fifth Avenue, which should remain a relatively sunlit shopping street, in the Times Square theater district, and at the southern edge of Central Park. Located around these, related to transit, would be the extensions of established office areas (such as Rackefeller Center and Penn Station) and new office centers. Housing would rise in the four corners, with parks along the river.

TRANSLATION

曼哈顿中城的办公空间预测

曼哈顿中城应该为更多的办公建筑建设做好准备,在接下来的35年里,(这些办公建筑的)建设量相当于大约66座时间人寿大厦,8个洛克菲勒中心或8个世界贸易中心。这些办公建筑提供了40万个工作岗位,占整个曼哈顿中央商务区预计增加就业人数的80%。

此项建议中包含的主要因素有以下几点。

①该地区的经济预测显示,纽约研究区的31个县的办公楼约增加了140万个工作岗位。(区域规划协会,地区的增长,1967年5月)

②一项调查发现,在该地区预计的就业岗位总数中,很大一部分将是企业总部,企业的类型与大型组织的商业政策相关。这在历史上的曼哈顿CBD的发展中很盛行。

③前面章节讲到的,通过谨慎应用功能性、形式性和适用性原则,使曼哈顿中城在不牺牲效率、便利性和环境质量的前提下,将容纳下这40万名

员工变成可能。

④对于在CBD之外无法妥善处理大量办公建筑的情况,将根据该地区的社会需求在老城区保留大量工作岗位。(区域规划协会,牙买加中心,1988年4月)

由于一个服务业的工作岗位通常需要支持到五个办公型的工作岗位,因此大约需要提供80,000名额外的工作人员。如果该地区有因商业、购物或娱乐目的而来的访客,该数据也会增加。

预计办公楼的增长

曼哈顿中城需要可容纳400,000名办公人员的办公空间,面积约8000万平方英尺的可租赁空间,也就是相当于66座时间人寿大厦的空间大小。这里选择时间人寿大厦来说明曼哈顿中城区办公建筑预计的扩建规模。它建于1959年,共有48层,在1,500,000平方英尺的可出租办公空间中雇用了6,000多名员工。这远远超过战后的"平均"办公楼尺度(平均层数30层楼,有些拥有超过500,000平方英尺的可出租空间),但这里选择它是因为它独特的尺度和覆盖范围(大半个城市街区)已经成为了办公空间的象征。当然,未来的建筑物的大小会有所不同,我们既不期望也没想要在地图上把他们的位置分布现在标注出来。

中城商业区中的永久性区域
目前最有可能使用的场所和不应该考虑的潜在办公场所

鉴于新办公楼的数量庞大,它们应该被安放在哪里?上面概述的是中城区,目前已成为容纳大型办公空间的区域。所有这些领域现在都在使用,但有些结构显然比其他结构更能抵抗变化,有些结构因社会和舒适性原因应该更能抵抗变化。地图上的所有彩色地块表示的是该地块被相对更为永久性的用途占用。图例中显示的前四个类别是相当客观标准的"硬空间":所有1948年后的建筑预计将持续到本世纪末期;一些重要的战前建筑,如洛克菲勒中心、克莱斯勒大厦和类似建筑也包括在此类别中。客观上重要的购物区也是"硬空间",如第五大道和先驱广场,以及主要机构和官方指定的地标。图例中的最后两个类别是由于主观原因和设计原因而形成的"硬空间"。对于商业娱乐区的剧院、俱乐部、酒店以及有成为地标可能的特色商店和私密餐馆,尽管它们可能在办公区中难以安置,并且在其他用途区中具有更大的经济价值,但在办公区中也仍然具有保留价值。

市中心商业区的"硬"与"软"区
与办公楼一起发展的永久性区域

在与上一幅图片(见前文图)相对应的地图中,没有着色的区域中的地块被认为是暂时性的或"软的"。这些地方通常最容易开发;在上一张地图上,它们以蓝色显示。正如上面的红色数量所证明的那样,中城的大部分用途似乎是或应该是相对永久的。但是有很多土地用途也可能会发生变化。从理论上讲,区域内应有足够的办公用地,以适应预计的城市扩张。然而,在实践中,从目前的房地产发展现状上看,上面显示的所有"软空间"中不能也不应该布置办公楼。虽然要保持高低对比的城市设计目标,但某些区域,尤其是剧院区域,尽管它们包含许多"软空间",但仍应保持"低"的空间形态。此外,如地图所示,大部分"软空间"的用地都不靠近地铁站。应当优先发展通向"软空间"的明确路径,创造更便捷的到达方式。

作为形式决定因素的可及性
地铁车站短距离内的"SOFT"区域

数千人工作的大型建筑物应位于铁路和地铁站的步行范围内。上面的地图展示了地铁站步行三分钟内的区域范围和"软空间"与可开发地块的范围。由此产生的重叠区域是目前在办公室区内可划为正在形成的"便捷路径"的区域。这些也是实施开发控制的优先站点,如第 5 章所述。"便捷路径"适用于中短程,但这些短途路径无法满足所有预计的开发。对于更大的范围,必须从包括目前划定的中城办公区以外的区域的更大范围考虑。

土地组合作为形式决定因素

公有土地所有权

目前划定的中城区商业区内的土地所有权是高度细分化的。与公众所熟知的不同,很少有半英亩或更多的"软空间"土地是单一所有权的。如上图(见前文图)所示,单一所有权中最大的可开发地块集中在哈德逊河西岸。以此为背景,办公区向哈德逊河的长距离扩张值得我们关注。此外,未来中城 CBD 的向西扩张在交通方面也是合乎逻辑的。任何 CBD 就业的实质性增长都需要超出现阶段的新快速交通的发展。鉴于曼哈顿的区域位置,通过市中心的最短路径是沿东西方向的。位于市中心的一条新的高速东西线不仅可以促进城市之间的联系,也可以刺激滨水区的商业开发,还可以将住宅开发重点放在曼哈顿东部和西部附近的未得到充分利用的地区。

概念图

中城的概念是从第 3 章讨论的功能和视觉形式原则发展而来的,并参考了前几页介绍的土地利用和运输分析。高层的办公塔楼集中于交通方式的关键节点上,被餐馆、专卖店、剧院、俱乐部和百货商店等的低密度区所分割。

图中显示了现有办公室集群和可能的新集群位置。现有办公室的"高点"为大中央车站、公园大道地带、洛克菲勒中心和宾州车站。可能的新办公楼群将包括西侧的洛克菲勒中心,向西扩展到第八大道以及宾州车站区域。

在中城的四个角落曾提出了高密度住宅群的发展模式。现在有向这种模式发展的趋势。

公共开放空间包围了哈德逊河上的客运站,并延伸到两条河流外围的高速公路上。

概念图只是迈向卓越城市设计的一步。它为许多小的决策提供了一个形式框架,这些决策可能会成为一个有意义的视觉整体。

城市形式和空间的构成组织不是这一概念的目标。其主要目标是实现视线的清晰,维持城市秩序和保证最大适宜性。

"条文"应限制在第五大道的建设,该大道应该是一条阳光相对充足的购物街,位于时代广场剧院区和中央公园的南部边缘。围绕着这个区域,坐落的是与运输相关的办公区域(如洛克菲勒中心和宾州车站)和扩建的新办公中心。在曼哈顿四个角落上的住房品质应有所提升,沿河有公园。

VOCABULARY

- equivalent [ɪ'kwɪvələnt] *adj.* 等值的;相等的;等同的
- underlying [ˌʌndə'laɪɪŋ] *adj.* 暗含的;潜在的;以……为基础的
- recommendation [ˌrekəmen'deɪʃən] *n.* 推荐,介绍;提议,建议;意见
- county ['kaʊnti] *n.* 郡,县
- headquarters [ˌhed'kwɔːtəz] *n.* 总部;总局;司令部
- amenity [ə'miːnəti] *n.* 生活福利设施,便利设施;娱乐消遣设施
- accommodation [əˌkɒmə'deɪʃən] *n.* 住处;工作场所;停留处
- tremendous [trɪ'mendəs] *adj.* 巨大的;极好的
- rentable ['rentəbl] *adj.* 可租的
- magnitude ['mægnɪtjuːd] *n.* 巨大;重大,重要性
- substantially [səb'stænʃəli] *adv.* 在很大程度上;大体上;基本上
- diagrammatic [ˌdaɪəgrə'mætɪk] *adj.* 图解的;概略的;图表的

- structure ['strʌktʃər] n. 结构;构造
- resistant [rɪ'zɪstənt] adj. 抵制的,阻止的,抗拒的
- parcel ['pɑːsəl] n. 一块土地;包裹,邮包
- occupied ['ɒkjəpaɪd] adj. 有人居住的;使用中的,有人使用的
- legend ['ledʒənd] n. 传说,传奇故事;民间故事;传奇人物;说明,图例
- criterion [kraɪ'tɪəriən] n. 标准,准则
- comprise [kəm'praɪz] vt. 包含;包括;构成,组成
- intimate-scale buildings 私密性建筑物
- susceptibility [səˌseptə'bɪləti] n. 敏感性,易受影响;敏感的感情,容易受伤的感情
- displacement [dɪ'spleɪsmənt] n. 被迫移居他乡;排水量
- impermanent [ɪm'pɜːmənənt] adj. 暂时的,短暂性的
- proximity [prɒk'sɪməti] n. 接近,邻近;临近
- overlay [ˌəʊvə'leɪ] vt. 在……上铺;掺杂着,交织着
- overlap [ˌəʊvə'læp] n. 重叠部分;相同之处
- implementation [ˌɪmplɪmen'teɪʃn] n. 实现;履行;安装启用
- concentration [ˌkɒnsən'treɪʃn] n. 专注,专心;集中;聚集;汇集;浓度
- merit ['merɪt] vt. 值得,应受
- increment ['ɪŋkrəmənt] n. 增加量,增长量
- underutilize [ˌʌndə'juːtɪlaɪz] vt. 未充分使用
- envelope ['envələʊp] n. 信封
- terminal ['tɜːmɪnəl] n. 站台;候机楼,航站;码头;计算机终端
- peripheral [pə'rɪfərəl] adj. 次要的,附带的,外围的,边缘的
- framework ['freɪmwɜːk] n. 架构,框架;体系,结构
- clarity ['klærəti] n. 清楚明了;清晰易懂;清晰;清楚

EXERCISES

[Choose the right answers]

1. With the projected office growth, why does the author mention the Time and Life building? _____.

   A. Because he wants to use the space of Time and Life building to describe the size of new office space needed in Mid-town Manhattan

   B. Because the Time and Life building is a magnitude project in Manhattan

   C. Because the Time and Life building is a representative of future buildings

   D. Because the Time and Life building has a long history since 1959

2. It can be inferred from the first map, the "hard" parcels _____.

   A. are all colored with white parcels on the map

   B. represent the area which can maintain for a long time

   C. are generally the most available place for development

   D. cannot be assembled for office buildings

3. We can infer from the first map, the "soft" parcels _____.

   A. are easily to be constructed

   B. seem to be or ought to be relatively permanent

   C. are comprised of uses

   D. contain permanent uses

4. According to the passage about "access trees", which following statement is not true?

   A. "Access trees" is the overlap areas within a three-minute walk of subway stations and "soft" or developable parcels.

B. "Access trees" are also priority sites for the implementation of development controls.

C. "Access trees" are not adequate for the long range.

D. "Access trees" can cover all of the projected development.

5. According to the passage, we can infer that _____.

A. clusters of high office towers centered on transportation access points separate "lows" such as restaurants, specialty shops, theaters, clubs and department stores are located

B. at the four corners of Mid-town high-density development patterns have been formed for years

C. public open space includes the superliner terminal on the Hudson, and extends over and beyond the peripheral highways on both rivers

D. the concept's main target is compositional organization

[Speak your mind]

1. How much office space does Manhattan need to provide?

2. In order to deal with the problem that most soft areas are not close to the subway station, what should we do?

3. What are the main characteristics of the land ownership in the current business district of Midtown?

Unit 13　The European Spatial Development Perspective

This chapter was published in its original form as:

Andreas Faludi(2004),Spatial Planning Traditions in Europe: Their Role in the ESDP Process. International Planning Studies,Vol. 9,155 - 172.

TEXT

In May 1999, in Potsdam, the ESDP received the blessing of the ministers of the member states of the EU responsible for spatial planning.

The document comprises two parts: a policy-oriented Part A and an analytical Part B. Part A starts with an introduction, "The Spatial Approach at European Level", asserting territory to be a new dimension of European policy. The opening sentence addresses a key concern as regards Europeanization, namely that it leads to more uniformity: "Spatial development policies... must not standardise local and regional identities... which help enrich the quality of life of its citizens". The spatial approach as such is about coordinating policies with a spatial impact. What is important is a shared discourse. Of course, there is reference to sustainable development. Here, the notion of balanced spatial development alluded to in the subtitle of the document: "Towards Balanced and Sustainable Development of the Territory of the EU" comes in. Balanced development helps to reconcile social and economic claims on land with an area's ecological and cultural functions. The medium through which this is to be done is that of a balanced spatial structure. Here, the document relates the three main policy guidelines of the ESDP:

①development of a balanced and polycentric urban system and new urban-rural relationship;

②securing parity of access to infrastructure and knowledge;

③sustainable development, prudent management and protection of nature and cultural heritage.

These guidelines must be reconciled with each other, and have due regard to local situations. However, the ESDP is no blueprint but rather "... a general source of reference for actions with a spatial impact... Beyond that, it should act as a positive signal for broad public participation in the political debate on decisions at European level and their impact on cities and regions in the EU". So the ESDP is a non-binding policy framework, and "... each country will take it forward according to the extent it wishes to take account of European spatial development aspects in its national policies".

Chapter 2 is about EU policies with a spatial impact that the ESDP wants to coordinate. The Single Market assumes space to be frictionless. This abstract idea is being imposed on a situation marked by long distances and physical barriers exacerbating cultural and linguistic diversity and different levels of development. The gains of integration are unevenly distributed. The Treaty of Rome recognizes this. The preamble refers to the need to "... reduce the differences between the various regions and the backwardness of the less favoured regions. In addition, some of the sectoral policies...assumed a regional character in their early phases". Some regulations were derogated to allow regions to catch up, but it took until 1975 when the European Regional Development Fund (ERDF) was established for any positive policy to emerge.

Regional policy is now the second-largest spender, after the Common Agricultural Policy, and a prominent examples of a Community policy with a spatial impact, and is also the cradle of European spatial planning. Originally a club of member states, through its regional policy the EU draws others levels of government within its orbit. The involvement of sub-national governments makes it a form of "multi-level governance".

Environmental policy is another EU spatial policy. The decision to embark on environmental policy was taken at the Paris Summit of 1972. The environment now has a prominent position in the European treaties. Most significant for spatial planning is the requirement of Environmental Impact Assessments and of setting aside areas for protecting birds and the habitats of endangered species, but the EU is also seeking to inject environmental awareness into all its policies.

The Treaty of Rome already asked for a transport policy, but it took a ruling of the European Court of Justice to jolt the Council of Ministers into action. Lobbying by the European Round Table of leading industrialists added to the momentum that led to the creation of so-called Trans-European Networks. The reasoning was that national networks needed to be integrated and access to them improved so as to facilitate the operation of the Single Market. Coordinating transport policy with spatial development policy and urban development measures would facilitate the desired shift in the modal split towards more environmentally friendly modes of transport, this being an example of where the "spatial approach" would come into its own.

The ESDP discusses other spatial policies as well, including competition policy and the policy of the European Investment Bank. Chapter 3 then presents 60 policy options, which constitute a mixed bag. Option 1 is "Strengthening of several larger zones of global economic integration in the EU through transnational spatial development strategies", contrast this with option 59 "Protection of contemporary buildings with high architectural quality". What is striking is the absence of any sort of key diagram conceptualizing European space—something that Dutch planners have pushed for. All that the ESDP gives is a verbal description of the core of Europe as the "pentagon" comprising London, Paris, Milan, Munich and Hamburg, the only "outstanding larger geographical zone of global economic integration" of Europe. Here, on 20% of the EU territory inhabited by 40% of its population no less than 50% of the EU's total GDP is being generated. The trend towards more concentration needs to be broken: "The creation of several dynamic zones of global integration, well distributed throughout the EU territory and comprising a network of internationally accessible metropolitan regions and their linked hinterland... will play a key role in improving spatial balance in Europe". The emphasis is on initiatives from below, which demonstrates the affinity of this concept with the notion of endogenous development and of building social capital (as one of the planks of EU regional policy).

This is indeed a central proposition, but one that is hidden behind a plethora of other concerns for which the compromise character of the ESDP, creating a need to accommodate many concerns, is responsible.

Chapter 4, on the application of the ESDP, specifies the intended follow-ups, all of which are on a voluntary basis. There are recommendations concerning the European Spatial Planning Observation Network (ESPON). A whole paragraph is devoted to transnational cooperation, endorsing the so-called INTERREG Community initiative. A further paragraph stresses the need for the Europeanization of state, regional and urban planning.

Chapter 5 of the ESDP is a perfunctory exploration of the impact of enlargement, then still way off but now, of course, a fact. The ESDP has come about by way of protracted discussions between national planners extending over the best part of 10 years. National establishments perceived opportunities to improve their positions; the process has been described elsewhere. The trigger had been the reform of the Structural Funds. The Directorate-General XVI responsible for such matters wanted to explore the spatial dimension of these vastly increased funds. To this end, Article 10 of the new regulations governing the ERDF was invoked for a study aiming to identify the elements necessary to establish a prospective outline of the utilization of the Community territory. On this basis, Directorate-General XVI produced "Europe 2000", followed later by "Europe 2000".

The end of the 1980s was also when ministers from the Member States gathered at Nantes in France for the first of what was to become a series of meetings on the road to Potsdam 10 years later. Ministers listened to Commission President Jacques Delors' views on cohesion policy. The Italians organized a follow-up. This suited the Dutch who, during their Presidency in 1991, wanted to hold another meeting. They took a great leap forward by proposing setting up the Committee on

Spatial Development (CSD) to prepare future meetings. Normally, the Commission would chair such a committee, but the proposal was for the rotating EU Presidency to hold the chair. German misgivings about the whole undertaking (about which more below) had been conveyed to other member states, which is why the Commission was held at a distance, but even so Directorate-General XVI provided the secretariat of the CSD. It gave other support as well, thereby treating the CSD as one of the untold number of Brussels committees.

At the fourth meeting in Lisbon various ministers asked for a spatial vision, and the next meeting at Liège agreed to the making of a Schéma de développement de l'espace communautaire (European Spatial Development Perspective) as the intergovernmental complement to "Europe 2000". In fact, the Germans already had plans for such a document to be prepared during their upcoming term.

In this they succeeded to the extent of getting the so-called Leipzig Principles, a kind of constitutive ESDP document, adopted in 1994. After Leipzig the participants expected to proceed swiftly. However, during their Presidency the French introduced various scenarios. The effort was inconclusive. Being next in line, Spain feared that the ESDP might imply reallocation of the Structural Funds. The Spanish central authorities were not enamoured by local and regional empowerment either.

Such ambivalence continued during the Italian Presidency. Nevertheless, ministers resolved to bring the process to its conclusion under the Dutch Presidency in 1997, which is when diagrammatic representations became an issue. Being unfamiliar with maps as a way of articulating spatial policy, some delegations were unhappy, and so the policy maps pushed for by the Dutch were relegated to the appendix.

At Noordwijk in the Netherlands, ministers accepted the "First Official Draft" of the ESDP. Transnational seminars and consultations within the Commission followed. In parallel, the British Presidency produced the "First Complete Draft".

Another German Presidency brought the process to a conclusion (but rescinded policy maps altogether). There was no fanfare, just a communique by the German Presidency announcing that the political debate had come to an end. In October 1999, the Finnish Presidency held a follow-up meeting at Tampere devoted mainly to a 12-point Action Programme.

The member states and the Commission are now applying the ESDP. An important arena is the INTERREG Community Initiative providing co-financing for hundreds of transnational projects. There is also ESPON ostensibly doing the groundwork for the next ESDP. We now turn to the motivations of the main players, which is where spatial planning traditions come in.

TRANSLATION

欧洲的空间发展观

1999年5月，ESDP在波茨坦得到负责空间规划的欧盟成员国部长的批准。

该文件包括两部分：政策导向部分A和分析部分B。A部分首先介绍"欧洲层面的空间方法"，主张领土是欧洲政策的新维度。开头句提到一个欧洲一体化的关键问题，即它会导致更强的一致性："空间发展政策……不能将地方和区域的特点变成标准化……应当帮助市民提升生活质量。"这样的空间方法是从空间角度出发的协调政策。重要的是共享的讨论。当然,也涉及可持续发展。在这里,文件副标题提到了平衡空间发展的概念:"实现欧盟领土的平衡和可持续发展。"平衡发展有助于协调区域内社会经济诉求与本地区的生态文化功能。实现这一点的媒介物是平衡的空间结构。在此,文件涉及 ESDP 的三个主要政策方针:

①平衡化开发、多中心的城市体系和新型城乡关系;

②确保获得基础设施和知识的机会均等;

③可持续发展、审慎管理和保护自然和文化遗产。

这些指导方针必须彼此协调,并适当考虑当地情况。然而,ESDP 并没有一个具体蓝图,只是提出"……提供一个通用来源为空间影响行为提供参考……除此之外,应当作为一个促进公众广泛参与的积极信号,在那些形成欧洲层面决定并影响欧盟各个城市和地区的政治辩论中"。因此,ESDP 是一个不具备约束力的政策框架,并且"……每个国家将依据 ESDP 所期望的外延来将之向前推进,将欧洲空间发展方面内容融入国家政策"。

第二章是关于 ESDP 希望协调的具有空间影响的欧盟政策。单一市场假定空间无摩擦。这种抽象的想法正被强加于另一种情况,这种情况的特征是长距离和物理障碍加剧了文化和语言的多样性以及发展水平不同。区域一体化的收益分布是不均匀的。《罗马条约》承认这一点。序言提到有必要"……减少各地区之间的差异和欠收益地区的后退倾向。此外,一些部门性政策……在早期阶段就假定了某一区域的特点"。一些规定被废除,以允许各地区向前追赶,但直到 1975 年,欧洲区域发展基金(ERDF)才被建立并推出了一些积极的政策。

区域政策现在是仅次于公共农业政策的第二大支出项,也是具有空间经济效果的公共政策突出的例子,并且也是欧洲空间规划的发源地。欧盟最初是一个成员国俱乐部,通过其区域政策吸引其他级别的政府加入。地方政府的参与使其成为一种"多层次治理"。

环境政策是欧盟的另一项空间政策。开始执行环境政策的决定是在

1972年巴黎首脑会议上做出的。现在环境政策在欧洲条约中占有突出的地位。对空间规划来说，最重要的是进行环境影响评估，并为保护鸟类和濒危物种的栖息地留出区域，但欧盟也在想办法将环境意识纳入其所有政策。

《罗马条约》已经要求制定交通政策，但欧洲法院做出了一项裁决，促使部长理事会采取行动。欧洲主要工业家圆桌会议的游说为所谓的"跨欧洲网络"的诞生增添了动力。理由是，国家网络需要加以整合，并改善对它们的访问，以便单一市场的运作。将运输政策与空间发展政策和城市发展措施协调起来，将有助于实现理想的模式划分向更有利于环境的运输方式转变，这是"空间方法"获得承认的一个例子。

ESDP还讨论了其他的空间政策，包括竞争政策和欧洲投资银行的政策。第3章提出了60个政策选项，它们构成了一个混合包。选项1是"通过跨国空间发展战略，加强欧盟几个较大的全球经济一体化区域"，与选项59"保护具有高建筑质量的当代建筑"形式对比。引人注目的是没有任何形式的关键图表来概念化欧洲，荷兰规划者全力推动ESDP，给出口头描述：欧洲的核心是"五角大楼"，包括伦敦、巴黎、米兰、慕尼黑和汉堡，欧洲是仅有的"全球经济一体化的杰出的大型地理区域"。在这些地方，20%的欧盟领土上居住着40%的人口，且有着不少于50%的欧盟GDP总量。需要打破这种愈发集中的趋势："在整个欧盟领土上建立几个动态的全球一体化区域，并由国际可及的大都市区域及其相连的腹地组成网络……将在改善欧洲的空间平衡方面发挥关键作用。"重点是来自下面的倡议，这表明这一概念与内部发展和建立社会资本（作为欧盟区域政策的一个支柱）的概念密切相关。

这确实是一个中心命题，但它隐藏在其他的大量问题的背后，而ESDP的妥协特性导致了许多问题，其应担负责任。关于ESDP的应用，第4章规定了预期的后续行动，所有这些都是在自愿的基础上的。这里有关于欧洲空间规划观测网（ESPON）的建议。整个段落着重讲述跨国合作,赞同所谓的国际社区倡议。另一段则强调需要将国家、区域和城市规划欧洲化。

ESDP 的第 5 章是对扩大的影响的表面探索,无论当时还是现在,这都是一个事实。ESDP 是通过 10 年来最好的国家规划者之间的长期讨论才产生的。国家机构认为机会提升其地位;这个过程已经在其他地方被描述过了。触发因素是结构性基金的改革。负责这类事务的总理事会第十六次会议希望探讨这些大量增加的资金的空间维度。为了这个目的,在一项研究中援引了管理 ERDF 的新条例第 10 条,目的在于确定建立对社区领土利用的前瞻性大纲所需的要素。在此基础上,第十六任总干事创作了"欧洲 2000",之后又再版了"欧洲 2000"。

20 世纪 80 年代末,欧盟成员国的部长们也齐聚法国南特,参加在此后十年通往波茨坦的道路上举行的一系列会议的第一次会议。部长们听取了委员会主席雅克·德洛尔关于凝聚力政策的看法。意大利人组织了后续行动。这对于在 1991 年担任主席的想再召开一次会议的荷兰人来说是合适的。他们提出设立空间发展委员会(CSD)以筹备今后的会议,这是向前迈出的一大步。通常情况下,委员会将主持这样的组织,但该委员会主席是由欧盟轮值主席担任的。德国人对整个承诺(下文更详细)的疑虑已经传达给其他成员国,这就是为什么委员会与其他成员国保持距离,但即使如此,总干事十六却向 CSD 提供秘书处。它也给予了其他支持,因此将 CSD 视为为数不多的布鲁塞尔委员会之一。

在里斯本的第四次会议上,各部长要求提出空间的愿景,下一个在 Lie'geagreed 举办的会议根据 Sche'ma de de'veloppement de l' 提出的空间法规(欧洲空间发展的角度)由政府来补充"欧洲 2000"。事实上,德国已经计划在即将到来的任期内准备这样一份文件。

在这方面,他们成功地在 1994 年通过了所谓的莱比锡原则,一种 ESDP 基本文件。在莱比锡事件之后,参与者们希望能迅速行动起来。然而,在他们担任总统期间,法国提出了各种设想。这项努力尚无定论。作为下一个目标,西班牙担心 ESDP 可能意味着结构性基金的重新分配。西班牙中央政府也不喜欢地方和区域授权。

这种矛盾情绪在意大利总统任期内依然存在。尽管如此,部长们还是决定在 1997 年荷兰总统任期内结束这一进程,当时图表上的陈述成为了一个争论点。一些代表团不满意用不熟悉地图作为空间政策的一种表达方式,因此荷兰推动的政策地图被归入附录。

在荷兰的诺德韦克,部长们接受了 ESDP 的"第一份正式草案"。随后在委员会内举行了跨国协商研讨会。与此同时,英国总统起草了"第一份完整草案"。

另一位德国总统结束了这一进程(完全废除了政策地图)。没有大张旗鼓,仅仅是德国总统宣布政治辩论已经结束的公报。1999 年 10 月,芬兰总统在坦佩雷举行了后续会议,主要讨论了 12 点行动方案。

会员国和委员会现在正在应用 ESDP。一个重要的领域是国际社区倡议,为数百个跨国项目提供共同融资。表面上看,ESPON 还在为下一个 ESDP 做基础工作。现在我们转向主要参与者,这就是空间规划传统的来源。

VOCABULARY

- policy [ˈpɒləsi] n. 政策,方针,策略
- dimension [dɪˈmenʃən] n. 空间,尺寸;层面,维度;特点
- sustainable [səˈsteɪnəbəl] adj. 可持续的,能长期保持的
- reconcile [ˈrekənsaɪl] vt. 调和;调解;使一致
- medium [ˈmiːdiəm] n. 媒介;方法,手段
- guideline [ˈgaɪdlaɪn] n. 指导方针,指导原则,准则
- heritage [ˈherɪtɪdʒ] n. 遗产
- blueprint [ˈbluːprɪnt] n. 蓝图,早期计划;早期设计

- frictionless [ˈfrɪkʃənlɪs] *adj.* 无摩擦的
- cradle [ˈkreɪdəl] *n.* 摇床；摇篮；传输座
- summit [ˈsʌmɪt] *n.* 峰会，首脑会议，最高级会议
- endangered [ɪnˈdeɪndʒəd] *adj.* 濒临灭绝的；有生命危险的
- lobby [ˈlɒbi] *v.* 游说，试图说服
- integrated [ˈɪntɪgreɪtɪd] *adj.* 综合的；完整的；互相协调的
- operation [ˌɒpərˈeɪʃən] *n.* 运作，实施，工作
- strengthen [ˈstreŋθən] *v.* 增强，加强，巩固
- transnational [ˌtrænzˈnæʃənəl] *adj.* 多国的；跨国的
- territory [ˈterɪtəri] *n.* 领土；领域，范围
- initiative [ɪˈnɪʃətɪv] *n.* 倡议；新措施；主动性；自主决断行事的能力
- trend [trend] *n.* 趋势，趋向；倾向，动向
- compromise [ˈkɒmprəmaɪz] *n.* 妥协；折中；让步；和解
- exploration [ˌekspləˈreɪʃən] *n.* 探测；勘查；探索；研究
- prospective [prəˈspektɪv] *adj.* 预期的；未来的
- secretariat [ˌsekrəˈteərɪət] *n.* 秘书处（人员）
- inconclusive [ˌɪnkənˈkluːsɪv] *adj.* 无结果的；无定论的
- reallocation [ˌriːˈæləˈkeʃn] *n.* 再分配
- appendix [əˈpendɪks] *n.* 附录
- ambivalent [æmˈbɪvələnt] *n.* 矛盾的；模棱两可的，含糊不定的

EXERCISES

[Choose the right answers]

1. According to the the article, the ESDP's main policy guidelines don't include _____.

A. new urban-rural relationship

B. sustainable development

C. single center city system

D. protection of natural cultural heritage

2. The main contents of part A of the document don't include _____.

A. sustainable development

B. that european territory should expand rapidly

C. that territory is a new dimension of European policy

D. the Spatial Approach at European Level

3. What's true about ESDP is that _____.

A. the ESPD has a blueprint

B. the ESPD wants to coordinate EU policies

C. the ESPD is a binding policy framework

D. the ESPD guidelines are coordinated but do not take into account local conditions

4. Regarding European space policy, _____ was not mentioned.

A. transport policy   B. environmental policy

C. regional policy   D. territorial expansion policy

5. The following statement about environmental policy is incorrect: _____.

A. it's European space policy

B. it has an important place in the European treaty

C. it was proposed in 1972

D. the decision to implement it was made in Paris

6. Europe's core cities do not include _____.

A. Milan   B. Munich

  C. Hamburg   D. Edinburgh

  7. What's wrong with ESPD chapter 4? _____.

  A. Committed to transnational cooperation

  B. There are Suggestions on ESPON

  C. Anticipated follow-up action was set out

  D. Oppose international community initiatives

[Speak your mind]

  1. What policies do ESPD include?

  2. What do you think of the compromise feature of ESPD?

  3. Do you know anything else about ESPD? If you know, tell me about it.

Unit 14　Reinforce Hub Functions of Hong Kong

This chapter was published in its original form as:
Hong Kong Special Administrative Regional Government (2007), Hong Kong 2030 Planning Vision and Strategy. 181-186.

TEXT

Measure Ⅰ　Ensure Adequate Supply of Land for Grade A Offices at Strategic Locations

Adequate supply of land for Grade A offices at the Central Business District (CBD) is needed to support the finance and business services sectors and to enhance Hong Kong's attraction as a choice location for corporate headquarters functions. These activities have a tendency to agglomerate at the core business districts and demand the best locations in town. To ensure adequate supply of Grade A office land, especially land suitable for office development of the highest grade, we need to consolidate and upgrade the existing CBD on the one hand, and explore opportunities for the development of further prime-grade office nodes outside the CBD on the other. The specific objectives are:

①To allow the gradual release of planned office sites at the CBD in response to market demand;

②To rezone suitable "government, institution, community" sites at the CBD for office use;

③To facilitate further office development at Quarry Bay and West Kowloon, and reserve land in Kai Tak for the possible development of a high-quality office node in the longer term;

④ To facilitate redevelopment by the private sector for the production of Grade A offices;

⑤ To encourage decentralisation of office activities to secondary office nodes at the urban fringe (e.g. former industrial areas) to facilitate a filtering process;

⑥To free up spaces at private office buildings at the CBD currently occupied by government uses;

⑦To enhance public transport accessibility, emphasise high-quality urban design and improve the public realm of the existing CBD.

Measure II   Facilitate Development for General Business Uses

In recognition of the increasingly obscured distinction between industrial and office (other than CBD Grade A offices) uses in Hong Kong, a new land use typology has been proposed for the purpose of forecasting future floor space demand—"General Business".

Currently, about 200 hectares of former industrial land have been rezoned to "OU(B)". These are mainly located at the fringe of the Metro Area and have good mass transport connections. We will need to continue to monitor future supply and demand of both business and industrial uses and allow for further rezoning when necessary.

Measure III   Provide Flexibility for Special Industries

While low-value-added industrial uses tend to fade out from Hong Kong as they gradually move across the boundary, we still have a thriving industrial sector which is sustained through increasing labour and floorspace productivity. It is primarily in the technology, media and other service sectors that productivity changes have occurred in situ and it is here that high productivity space developments in Hong Kong such as the Science Park and Cyberport have been formed to help increase the added value and productivity of Hong Kong's industry. Similarly, the Applied Science and Technology Research Institute (ASTRI) is also geared to upgrading the technology level of our industry and stimulating growth of technology-based industry in Hong Kong. The Science Park has promoted development of four focused clusters in electronics, information technologies and telecommunications, precision engineering and biotechnology. ASTRI has focused on four technology areas, namely communications technologies, consumer electronics, integrated circuit design and opto-electronics.

Spatial requirements for high-value-added industrial processes, with emphasis on special accommodation requirements and tailor-made design, are very different from those for traditional low-value-added industries-those which conventional multi-storey factories would fail to meet. Such needs could probably be accommodated at industrial estates or special industrial zones.

Hence, to ensure a strong and diversified economic base for Hong Kong, we need a strategic reserve of land for special industries so that our infrastructure would not lag behind opportunities as and when they arise, and to cater for land-extensive operations or investments which are of major, if not strategic, importance to Hong Kong. As such, other than continuous revamping our existing facilities, i. e. the industrial estates, Science Park and Cyberport, we also need to seek out further opportunities for strategic land reserves suitable for special industries. We would recommend that these be incorporated as part of the NDA developments to be further studied.

Measure IV　Enhance the Port and Airport

Hong Kong's port and airport serve as important infrastructure in support of our role as an international and regional transport ion and logistics hub. We need to coordinate closely with the Port Development Strategy and Airport Masterplan, especially in enhancing the efficiency of the port and airport through improving the operation mode, enhancing connection with the hinterland, and advancing intra-regional cooperation. We need to monitor growth trends and plan well ahead for future airport and port expansions, if needed, to address possible future demand.

Measure V

We should continue to enhance our tourism infrastructure and make better use of our many resources (including cultural/heritage, marine and other natural resources) to diversify our offer. This could help us develop niche markets and promote the fast-growing sectors of alternative tourism such as ecotourism and cultural tourism. We should also continue to work with our Mainland and Macao counterparts through the Hong Kong Tourism Board for the promotion of more multi-destination travel in the PRD region.

TRANSLATION

加强香港的枢纽功能

措施一　确保在策略地点提供充足的甲级办公室用地

在商业中心区必须提供充足的甲级办公室用地,以支援金融及商贸服务行业,吸引机构在本港设立总部。这些业务集中在商业中心区,设置于城市的黄金地段。为确保甲级办公室用地的供应充足,尤其是适合作最优质办公室用地的发展,我们一方面需整合及提升现有的商业中心区,另一方面应开拓在商业中心区以外地点发展其他优质办公室枢纽的机会。具体目标如下:

①按市场需要,逐步释放出位于商业中心区的已规划的办公用地;

②改建一些商业中心区内的"政府、机构及社区"的用途地带作为办公室;

③在则鱼涌和西九龙进一步发展办公用地,并在启德预留土地,长远而言,可作为发展优质办公室枢纽之用;

④推动私人机构与甲级办公室的重建项目;

⑤鼓励一个市场筛选的过程,将不需要在商业中心区运作的办公室经济活动布置到市区边缘的次级办公地带(例如一些原有工业区);

⑥腾出现时处于商业中心区内政府租用的私人办公室;

⑦加强往来现有商业中心区公共交通的可达性,着重优化城市设计以及改善其公用地方环境。

措施二 加速一般商业用地的发展

鉴于香港的工业与办公用途的区别日渐模糊,我们在预测未来的楼面空间需求时,提出了一种新的土地用途类型,即"一般商业"用途。

目前,约有200公顷的前工业用地已被重新划定为"商贸"用地。这些用地主要位于都会区的边缘地带,有良好公共交通接驳。我们需要不断继续监测未来商业及工业用地的供求情况,并在有需要时进一步规划用途地带。

措施三 弹性地使用特殊工业用地

当低增值工业因逐渐通往华南地区而在香港逐渐淡出时,我们仍有一些因需要大量劳动力或需要庞大的场所而繁华的场地。生产力改变主要表现在本港的科技、传媒及其他服务行业。此外,香港科学院和数码港等高生产力空间发展项目,亦有助提升本港工业的附加值及生产力。同样地,政府还成立香港应用科技研究院以提升本港工业的技术水平和带动以科技为主的工业增长。科学园推动了四个重点领域,包括电子、讯息科技和电讯、精密工程和神股网科技。香港应用科技研究院则主要集中于四个技术范围,即通讯技术、消费电子、集成电路设计和光电子。

高增值工业与传统的低增值工业对空间的需求截然不同,其生产流程注重特殊的环境设置及合乎需要的设计场地。传统的多层工厂大厦未必能够满足其有关需求。因此高增值工业可能适用于工业区或特殊工业区。

因此,为了确保本港具有多元化而稳固的经济基础,我们必须策略性地预留土地作特殊工业用途,以便商机涌现时,我们能把握有关机遇,提供配套土地,以满足一些对本港十分重要(若非关键性)、广占用地的生产业务或投资的需要。为此,除了不断改善工业区、香港科学园以及码港等现有设施外,我们亦应寻求其他机会,策略性地预留土地作特殊工业用途。我们建议将这些土地纳入新发展区的发展计划,以待进一步研究。

措施四 提升港口及机场设施

本港的港口和机场是支持香港作为全球及区域运输及物流中心的重要基础设施。我们必须紧密配合港口发展策略及机场发展蓝图,尤其通过改善运作模式、加强与大陆的联系及促进区内机场合作等方面去提升港口及机场的效率。我们需要监测增长趋势,及早规划机场及港口的扩建计划(如需要扩建的话),以解决未来可能出现的需求。

措施五

我们应继续提升本港的旅游基础设施,并充分利用本港的丰富资源

（包括文化、文物古迹、海洋及其他自然资源），促进旅游业务多元化发展。此举将有助本港发展特色市场，并促进正在快速增长的生态旅游及文化旅游等另类的旅游业的发展。此外，我们还应继续通过香港旅游发展局与大陆及澳门携手合作，推动珠三角地区的一程多站式旅游。

VOCABULARY

- agglomerate [ə'glɒməreɪt] *adj.* 成团的，聚结的
- consolidate [kən'sɒlɪdeɪt] *vt.* 巩固；加强；合并，联合
- rezone [ˌriː'zon] *vt.* 再分区，重定分区用途
- node [nəʊd] *n.* 节，结，瘤；交点
- fringe [frɪndʒ] *n.* 边缘，外围
- filter ['fɪltə] *v.* 过滤，滤除
- realm [relm] *n.* 界，领域，范围；王国
- typology [taɪ'pɒlədʒi] *n.* 类型学
- forecast ['fɔːkɑːst] *vt.* 预报，预测
- conversion [kən'vɜːʃən] *n.* 改变，转变；转化
- flexibility [ˌfleksə'bɪləti] *n.* 灵活性；弹性；适应性
- obsolete [ˌɒbsəl'iːt] *adj.* 废弃的；过时的；淘汰的；老化的
- annotate ['ænəteɪt] *v.* 给……作注解；给……加评注
- boundary ['baʊndəri] *n.* 分界线；边界；界限
- labour ['leɪbər] *n.* 劳动；（尤指）体力劳动；工人，劳工
- situ ['sɪtʃuː] *n.* 原地
- telecommunication [ˌtelɪkəmjuːnɪ'keɪʃən] *n.* 远程通信；无线电通信
- precision [prɪ'sɪʒən] *n.* 准确，精确，精密；严谨；细致，认真
- strategic [strə'tiːdʒɪk] *adj.* 有助于计划成功的；战略的；战略性的

- cater for 迎合；供应伙食；为……提供所需
- revamp [ˌriːˈvæmp] vt. 修改；改进
- coordinate [kəʊˈɔːdɪneɪt] v. 协调；使相配合；相配，相称；协调
- hinterland [ˈhɪntəlænd] n. 内陆，内地；边远地区
- marine [məˈriːn] adj. 海的；海产的；海生的
- niche market 利基市场
- ecotourism [ˈiːkəʊˌtʊərɪzəm] n. 生态旅游
- mainland [ˈmeɪnlænd] n. 大陆；本土

EXERCISES

[Choose the right answers]

1. To ensure adequate supply of Grade A office land , we need to _____.

    A. upgrade the building of CBD district

    B. seek chances for the evolution of Grade A office nodes within the CBD

    C. pay more attention to the suitable land for office development of the highest grade

    D. enhance the building of existing CBD and extend the opportunities outside of the CBD

2. Why do we create the new type of land use which called "General Business"? _____.

    A. Because the clear boundaries between industrial and office are disappearing in recent years

    B. Because distinction between industrial and office have been founded

C. Because it can predict future floor space demand

D. Because the new uses are more popular in Hong Kong recently

3. The difference between high-value-added industries and low-value-added industries is _____.

A. the economic base

B. the spatial requirements

C. the cater for land-extensive

D. the operations or investments

4. Many measures can be taken to improve the efficiency of our port and airport except _____.

A. improving the operation mode

B. enhancing connection with the hinterland

C. advancing intra-regional cooperation

D. building an international and regional transport ion and logistics hub

5. Which is the best title for Measure V? _____.

A. Improve our Offer on Tourism

B. Develop Tourism Service in Hong Kong

C. Use New Transportation to Connect the Traffic

D. Maintain Our Tourism Facilitate

[Speak your mind]

1. Why do we need to provide sufficient land supply for Grade A office buildings in the Central Business District?

2. What is the main location of land that has been reclassified from industrial land to "OU(B)"?

3. What areas of science park have promoted development?

4. What are the critical infrastructures that support international and regional transportation and logistics?

5. What are the main measures to enhance the tourism brand?

Unit 15   Residence Districts in NYC Zoning

This chapter was published in its original form as:

New York City Department of City Planning (2011). Zoning Handbook. Chapter 3: Residence Districts.

TEXT

Residence districts are designated by the prefix R in the Zoning Resolution. There are ten standard residence districts in New York City—R1 through R10. The numbers refer to the permitted density (R1 having the lowest density; R10 the highest) and certain other controls such as required parking. A second letter or number signifies additional controls in certain districts. Unless otherwise stated, the regulations for each of the ten residence districts pertain to all sub-categories within that district. The R4 district, for example, encompasses R4-1, R4A and R4B.

Standard Districts

R1 and R2 districts allow only detached single-family residences and certain community facilities. The R3-2 through R10 districts accept all types of dwelling units and community facilities and are distinguished by differing bulk and density, height and setback, parking, and lot coverage or open space requirements.

R3-2 districts permit detached and semi-detached houses, garden apartments, rowhouse developments and a broader range of community facilities. R4 and R5 zones are primarily districts of rowhouses and small multiple dwellings. The R6 through R9 districts without a letter suffix (R8 rather than R8A, for example) encourage on-site open space and on-site parking. These objectives are addressed by a complex formula involving three variable controls: floor area ratio (FAR), height factor (HF), and open space ratio (OSR). The Zoning Resolution assigns a range of floor area ratios in these districts. The maximum floor area ratio in each district is reached for a building with a specific height factor in combination with a specific open space ratio often resulting in a tall, low-coverage building set back from the surrounding streets. Although there is no range of floor area ratios in R10 districts, the tower provisions and the 20 percent floor area bonus for plazas encourage high-rise, low-coverage buildings set back from the streets. This open space emphasis in R6 through R10 districts sometimes leads to the construction of buildings that are out-of-scale with the surrounding neighborhood, breaking the existing street wall continuity which characterizes many New York neighborhoods.

Contextual Districts

In 1984, 1987 and again in 1989, the Zoning Resolution was amended to establish a number of new and revised residential districts. These districts, generally identified with the suffix A, B, X or 1 (except R7), are termed contextual because they maintain the familiar built form and character of the existing community while providing appropriate development opportunities.

Lower Density Contextual Districts

In recent years, out-of-scale construction in low-rise neighborhoods had blurred the distinctions between residence districts. Sound one- and two-family houses were often demolished and replaced by larger, multifamily buildings. There was a need to determine regulations for appropriate new development in low-rise neighborhoods and to preserve existing housing. In 1989, New York City enacted the first comprehensive revision of lower-density zoning since 1961.

Lower density contextual zoning reaffirms the bulk distinctions, building configurations and narrower lot sizes of many older residential neighborhoods. By controlling curb cuts, it also provides more on-street parking and discourages excessive paving of front yards. It is applicable to low-rise neighborhoods in the Bronx, Brooklyn, Queens and Staten Island.

Six new contextual residence districts were created (R2X, R3A, R4-1, R4A, R4B, R5B) to recognize the particular characteristics of detached and semi-detached residence and rowhouse neighborhoods. One existing residence district (R3-1) was reconfigured as a contextual district and three other general residence districts (R3-2, R4, R5) were modified to incorporate elements of lower density contextual zoning.

New requirements were established to maintain the contextual cohesion of these new and amended districts. All usable living space, including most enclosed garage and attic space, must now be counted in floor area calculations. However, in R2X, R3, R4, R4-1 and R4A districts, a new attic allowance permits an increase in FAR for floor area under a pitched roof with headroom between five and eight feet. A new zoning envelope sets overall building heights for each district as well as a maximum perimeter wall height, above which pitched roofs or setbacks are required, to minimize the visual impact of new buildings on the street. Typically, R3, R4, R4-1 and R4A districts promote houses with pitched roofs while R4B and R5B zones are primarily rowhouse districts. Driveways that run parallel to the side lot line encourage traditional landscaped front yards and side yard parking in detached and semi-detached residence districts. Limitations on the width, number and location of curb cuts maximize on-street parking and lessen neighborhood parking problems.

Medium and Higher Density Contextual Districts

A major emphasis of the 1961 Zoning Resolution was the construction of tall, slender buildings surrounded by large, open spaces. However, new residential development was often incompatible with the character and configuration of older neighborhoods. The cost and inefficiencies associated with construction of these buildings contributed to a slowdown in housing production. In 1984 and 1987, the Zoning Resolution was amended to establish a number of contextual districts in medium and higher density residential areas (R6A, R6B, R7A, R7B, R7X, R8A, R8B, R8X, R9A, R9X, R10A).

Medium and higher density contextual districts combine maximum lot coverage with a requirement that buildings be placed on or near the street line and attain at least a certain minimum height within the street wall setback distance. In addition, front and rear sky exposure planes control the overall height of the buildings. Instead of a range of floor area ratios to be used in combination with various height factors and open space ratios, each medium and higher density contextual district allows the maximum floor area ratio on a zoning lot irrespective of height factor or open space ratio. The interaction of the floor area ratio, lot coverage and street wall requirements results in lower, bulkier buildings closer to the sidewalk that are in keeping with the scale and character of the existing neighborhood and which maintain the traditional streetscape. Because the tower provisions and plaza bonuses available in R10 districts are not permitted in R10A districts, new residential development will be similar in style to buildings in older, built-up neighborhoods.

Contextual and non-contextual districts treat non-residential and community facility buildings differently. In non-contextual districts, residential, non-residential and community facility buildings are subject to different floor area ratios and other bulk regulations. In medium and higher density contextual districts, the space requirements of community facilities can be adequately accommodated and, therefore, residential and community facility buildings are generally subject to similar bulk regulations. However, in lower density contextual districts, bulk regulations for residential and community facility buildings usually differ because community facilities are typically larger than residential buildings.

Contextual districts also differ from most non-contextual districts in the way they control the density of residential development. In the non-contextual R6 through R10 districts, density is measured in zoning rooms. Each zoning lot is restricted to a maximum number of zoning rooms. This number is determined by dividing the area of the zoning lot by the minimum number of square feet of lot area required for each zoning room. The minimum square feet of lot area required for each zoning room varies from district to district it also depends upon the floor area ratio, height factor and open space ratio used in the development In the contextual districts, density is measured in dwelling units. There is only one lot area requirement in each of these zoning districts. Each zoning lot is restricted to a maximum number of dwelling units, a number arrived at by dividing the area of the zoning lot by the minimum number of square feet of lot area required per dwelling unit. This allows more flexibility in laying out the interior of the dwelling units.

Quality Housing

As part of the 1987 amendments to the Zoning Resolution, the medium and higher density contextual district bulk regulations were made optional in corresponding non-contextual districts, and the Quality Housing Program was established as a mandatory requirement for all residential buildings developed under the medium and higher density contextual bulk regulations. The purpose of these amendments is to encourage development of multifamily housing in a way that recognizes the relationship between building design and the quality of life in a dense urban environment.

Under contextual lot coverage, the maximum floor area may be reached in a building with fewer stories than would be permitted under non-contextual zoning. For example, in an R7 district, under standard zoning, the maximum FAR of 3.44 is achieved only in a 14-story building. A six-story building would have an FAR of 2.88. However, under the contextual regulations, a six-story building could reach the full 3.44 FAR. In addition, in R6, R7 and R8 districts, on wide streets outside the Manhattan Core, buildings developed under the Quality Housing

Program may achieve slightly higher FARs.

The Quality Housing Program requires that all developments built under the medium and higher density contextual bulk regulations also comply with the four major elements of the Quality Housing Program: Neighborhood Impact, Recreation Space, Safety and Security, and Building Interior.

Neighborhood Impact is controlled primarily by the contextual bulk regulations outlined above, and by street tree planting and ground floor window requirements. Each of the other three program elements—Recreation Space, Safety and Security, and Building Interior—have several mandatory components and some of the components have a two-tier system of standards (minimum and preferred).

The Recreation Space element establishes minimum and preferred standards for the amount of equipped indoor and outdoor space, mandatory regulations for landscaping as a percentage of the open lot area, and on-site tree planting. If a development meets only the minimum standards for recreational space, instead of the preferred standard, it would have to meet the preferred level of compliance for the size of the average dwelling unit, a component of the Building Interior program. Other Building Interior requirements include windows larger than those required by the Building Code, laundry facilities and trash storage.

The Safety and Security element includes minimum and preferred standards for the number of apartments per corridor. Other requirements include: building entrances visible from the street, and elevators and stairs visible from both the building entrance and individual apartments.

TRANSLATION

纽约市居住区区划

居住区在区划决议中被指定为前缀 R。在纽约市有 10 种标准的居住区——R1 到 R10。数字代表许可的密度（R1 密度最低，R10 最高）和确定的其他控制条件，例如必需的停车位。第二个字母或者号码表示特定区域的附加控制条件。如果没有其他规定，每类居住区的规章适用于所有的子类别。例如 R4 区包含 R4-1，R4A 和 R4B。

标准区

R1 和 R2 区只允许独立式的单一家庭住宅和特定的社区设施。R3-2 到 R10 区接受所有类型的居住单元和社区设施，区别在于体积和密度、高度和退台、停车、地块覆盖率或者开敞空间需求（等）。

R3-2 区允许独立和半独立住宅、花园公寓、联排式住宅开发和更大范围的社区设施。R4 和 R5 区是联排住宅和一些多用户住房的主要类型区。没有字母后缀的 R6 到 R9 区（例如 R8 不包括 R8A）鼓励就地开敞空间和就地停车。这些目标通过一个包括三个变量的复合公式来实现：容积率、高度系数和开敞空间比率。区划决议赋予这些地块一定范围的容积率。每个地块的最大容积率达到了建筑的特定高度系数和特定开敞空间比率，常常使得高大的、低覆盖率的建筑从周边街道退让。即使在 R10 区没有容积率范围，塔楼条款和对广场的 20% 建筑面积补贴鼓励高层的、低覆盖率、退离街道的建筑。R6 到 R10 区强调的开敞空间有时导致建筑建造与周边街区不成比例，破坏现有的塑造了很多纽约街区的街墙连续性。

文脉区

1984 年、1987 年及 1989 年，区划决议被修订以建立大量新的和经过修订的居住区。这些区域通常用后缀 A，B. X 或 1 来标记（除 R7 以外），被称为文脉区,因为它们保持了现有社区的常见的建筑形式和特征,同时提供合适的发展机会。

低密度文脉区

近年,低层街区的不成比例建设使居住区之间的区别变模糊。完好的一户或两户家庭的房子经常被拆掉,被更大的多户建筑替代。需要确定规章来在低密度街区进行合理的建筑开发和现有建筑保护。在1989年,纽约市颁布了1961年以来第一个低密度区划的综合修订本。

低密度文脉区区划重申(重新肯定)了体量特质、建筑组合(方式)和很多历史居住街区的狭窄的地块尺度。通过控制路缘分割,也提供了更多路边停车位,阻止住宅前院的过度铺装。它适用于布朗克斯区、布鲁克林区、皇后区、史坦顿岛。

新增6类新的文脉居住区(R2X,R3A,R4-1,R4A,R4B,R5B)来识别独立、半独立住宅和联排住宅街区的特征。一类现有的居住区(R3-1)被修改为文脉区,3类其他普通居住区(R3-2,R4,R5)被修改以吸收低密度文脉区的元素。

建立新的要求以保持这些新改良区域的文脉凝聚力。所有便于使用的生活空间,包括最封闭的车库和顶楼空间,现在必须计入建筑面积。然而,在R2X,R3,R4,R4-1,R4A区,对于5~8英尺坡屋顶下的建筑面积,新的阁楼政策允许其增加建筑容积率。在需要坡屋顶和退台的前提下,一个新的区划限制为每个区设定(给出)了总体性的建筑高度和周边的墙高,来减少新建筑对街道的视觉影响。通常,R3,R4,R4-1和R4A区建设坡屋顶的独栋住宅,R4B和R5B主要是联排住宅。平行于地块侧面边线的私人车道鼓励传统园景前院和侧院,在独立和半独立居住区内停车。对路缘石宽度、数量和位置的限制,最大限度地提高了路边停车,减少了社区停车问题。

中高密度文脉区

1961年区划决议的一个要点是建设被大的开敞空间包围的高楼。然而，新的住宅开发经常与古老街区的特征和构型不协调。与这些楼宇建设相关的支出和低效导致了住宅建设的减速（衰退）。在1984年和1987年，区划决议被修订，创造了若干中高密度居住区的文脉区（R6A，R6B，R7A，R7X，R8A，R8B，R8X，R9A，R9X，R10A）。

中高密度文脉区具有最大的覆盖率，同时要求建筑位于或邻近道路边线，并且在沿街立面退线距离内达到一个最小高度。此外，前后天际面控制了建筑的高度。不再用三指标结合控制，每个中高密度文脉区允许规划地块的最大容积率，不管高度系数和开敞空间比率。容积率、地块覆盖率、沿街立面要求相结合，形成了一个低矮、庞大的建筑，更加接近人行道，与已有的街区尺度特征和传统街面景观一致。因为塔楼条款和广场奖励适用于R10区，但在R10A区不采用，开发的新的住宅会与旧的已建街坊在风格上相近。

文脉区和非文脉区对非居住建筑和社区服务设施建筑的要求不同。在非文脉区，居住、非居住和社区服务设施受到不同容积率和体量规定的管制。在中高密度文脉区，社区设施的空间要求可以被满足，因此居住建筑和社区设施通常遵从相同的体量规定。然而，在低密度文脉区，居住建筑和社区设施的体量条框通常是不同的，因为社区设施比居住建筑更大。

在控制居住开发密度的方法上，文脉区也不同于大多数的非文脉区。在非文脉区的R6-R10区，密度用规划房间数来衡量。每个规划地块受到一个最高规划房间数量的控制。这个数量等于规划地块面积除以规划地块每个房间需要的最小平方英尺数。规划地块每个房间需要的最小平方英尺数在各个区是不同的，取决于三大指标。在文脉区，密度用居住单元来衡量。在每个规划区内不只有一个地块面积要求。每个规划地块有一个最大的居住单元数量，等于规划地块面积除以规划地块每个居住单元需要的最小平方英尺数。在居住单元内部的布局方面，这个要求更加灵活。

优质住房

作为区划决议1987年修正案的一部分，对中高密度文脉区的体量控制在相应的非文脉区是非强制性的,优质住房项目被建立,作为一个对中高密度文脉体量控制规则下的所有居住建筑开发的强制性的要求。这项修改的目的在某种程度上是鼓励多家庭住房的开发,承认了密集城市环境中住房设计和生活质量之间的联系。

在文脉地块范围内,与非文脉区的区划相比,一栋建筑的最大建筑面积限制可能通过更少的层数。例如,在 R7 区,普通区划下,最高容积率 3.44 的实现只能在 14 层建筑。一个 6 层建筑的容积率是 2.88。然而,在文脉区规则下,6 层建筑可以达到最高容积率 3.44。另外,在 R6、R7、R8 区,在曼哈顿中心之外的宽阔街道上,在优质住房项目之下的建筑开发可以达到更高的容积率。

优质住房项目需要所有的在中高密度文脉体量规则下建设的开发项目也遵守优质住房项目的四点要素:邻里影响,休憩空间,安全和保安,建筑内部。

邻里影响主要通过上述文脉体量规则、道路树木种植、一层窗户的要求来控制。休憩空间、安全和保安、建筑内部这三个其他项目要素中的一个,有若干强制性的部分,一些部分具有双重的标准(最低标准和首选标准)。

休憩空间要素制订了对(需要)建设的室内和室外空间数量的最低和优先标准,对一定比例的地块开放区域进行景观美化的强制性标准,在地址上进行绿化种植(的标准)。如果一项开发只符合休憩空间的最低标准,而不是优先标准,它就必须符合平均居住单元尺寸的优先承诺,即建筑内部项目的一个组成部分。其他建筑内部要求包含窗户尺寸大于建筑规范要求、洗衣设施和垃圾堆放点。

安全和保安要素包括对每层房间数量的最少标准和优先标准。其他要求包括从街上可以看见建筑入口,从建筑入口和房间可以看见电梯和楼梯。

VOCABULARY

- residence ['rezɪdəns] *n.* 住所；住房；宅第
- designate ['dezɪgneɪt] *vt.* 指定，选定；委派；划定
- prefix ['priːfɪks] *n.* 前缀
- resolution [ˌrezə'luːʃən] *n.* 决议；正式决定
- pertain [pə'teɪn] *vi.* 关于，有关；
- sub-category 子类；分类；子类别；范畴
- detach [dɪ'tætʃ] *vt.* 使分离，使分开；拆掉
- dwelling ['dwelɪŋ] *n.* 房屋，住所，住处
- setback ['setbæk] *n.* 挫折；障碍
- semi-detached house 半独立屋
- rowhouse ['rəʊˌhaʊs] *n.* 排屋
- on-site [ˌɒn'saɪt] *adj.* 在工地的；在现场的
- formula ['fɔːmjələ] *n.* 惯例；配方；处方；公式，方程式；分子式
- variable ['veərɪəbəl] *n.* 变量 *adj.* 多变的；反复无常的
- assign [ə'saɪn] *vt.* 分配；分派；指派
- provision [prə'vɪʒən] *n.* 规定，条款
- bonus ['bəʊnəs] *n.* 奖金；红利；津贴；另外的优点；额外的好处
- continuity [ˌkɒntɪ'njuːəti] *n.* 连续性，持续性，连贯性
- amend [ə'mend] *vt. & vi.* 修订，修正，修改
- revise [rɪ'vaɪz] *vt.* 修订；修正；修改；复习；温习
- revised [rɪ'vaɪzd] *adj.* 修正过的；经过修改的
- blurred [blɜːrd] *adj.* 模糊的；弄不清的；分不清的；看不清楚的
- sound [saʊnd] *adj.* 无损伤的；完好的；健康的，健全的
- demolish [dɪ'mɒlɪʃ] *vt.* 拆除，拆毁；推翻，颠覆

- enact [ɪ'nækt] vt. 实行,实施;制定;上演;表演
- revision [rɪ'vɪʒən] n. 修订,修改;复习;温习
- reaffirm [ˌriːə'fɜːm] vt. 重申,再次声明;加强
- configuration [kənˌfɪgə'reɪʃən] n. 布局;构造,结构;格局;配置
- paving ['peɪvɪŋ] n. 铺过的地面;铺筑材料
- applicable [ə'plɪkəbəl] adj. 生效的;适合的,适用的
- reconfigure [ˌriːkən'fɪgər] v. 重新装配,改装
- modify ['mɒdɪfaɪ] vt. 修改,更改,改造,改变
- incorporate [ɪn'kɔːpəreɪt] vi. 包含;将……包括在内
- cohesion [kəʊ'hiːʒən] n. 凝聚(力);团结
- amend [ə'mend] v. 修订,修正,修改
- garage ['gærɪdʒ] n. 车库,汽车房;汽车修理厂;加油站
- attic ['ætɪk] n. 阁楼;顶楼
- pitched [pɪtʃt] adj. 倾斜的;定调的
- headroom ['hedruːm] n. 净空高度
- overall [ˌəʊvə'rɔːl] adj. 总的;全面的;包括一切的
- perimeter [pə'rɪmɪtər] n. 周边,边缘;周长
- driveway ['draɪvweɪ] n. 私人车道
- parallel ['pærəlel] adj. 平行的
- side lot line 侧面基地线
- lessen ['lesən] vt. 减少,降低,减轻
- slender ['slendər] adj. 苗条的,纤细的;少量的
- incompatible [ˌɪnkəm'pætəbəl] adj. 不相容的;不兼容的;不能和谐相处的
- street wall 沿街立面
- irrespective [ˌɪrɪ'spektɪv] adv. 不考虑地,不顾地
- bulkier ['bʌlkiə] adj. 庞大而占地方的

- amendment [əˈmendmənt] n. 修改,修订,修正;修正案,修正条款
- mandatory [ˈmændətəri] adj. 强制的;必须履行的;法定的
- tier [tɪər] n. 一层　v. 层叠;层层排列

EXERCISES

[Choose the right answers]

1. From the first paragraph, we can learn that _____.

A. R4 district must contain all the regulations which stipulate in the R4-1 district

B. R4-1 district must conform to the regulations which stipulate in the R4-A district

C. in most cases, R4-1 district should abide by the rules which regulated in R4 district

D. all sub-categories pertain to the regulations for each of the ten residence districts

2. By means of _____, the R6 through R9 districts without a letter suffix encourage on-site open space and on-site parking.

A. three variable controls in the formula

B. commanding the formula without a letter suffix

C. increasing FAR, HF, and OSR

D. decreasing floor area ratio, height factor, and open space ratio

3. According to contextual districts regulations, which one belongs to Contextual Districts?

A. R7A　　　　B. RA　　　　C. R13　　　　D. R1X

4. In R2X, R3, R4, R4-1 and R4A districts, a new attic allowance _____.

A. discounts most enclosed garage and attic space in floor area calculations

B. permits the increasing FAR for floor area

C. promotes houses with pitched roofs

D. encourages more curb cuts number limitations

5. The passage indicates that contextual districts and non-contextual districts _____.

A. have the same regulations of controlling the density of residential development

B. dispose non-residential and community facility buildings in different ways

C. are subject to same bulk regulations

D. are restricted to a maximum number of zoning rooms

[Speak your mind]

1. What kind of building facilities can we build in R1 and R2 districts?

2. What form do we usually use to mark the contextual districts?

3. In Lower Density Contextual Districts, what kind of usable living space new requirements established must be counted in floor area calculations?

4. In which districts the residential buildings and community facilities comply with the same volume regulations?

5. What is the purpose of the transition from Medium and Higher Density Contextual Districts to Quality Housing in the zoning resolution?

Reference Answer

Unit 1 Garden City

1. C; 2. C; 3. C; 4. B; 5. A

1. By being a healthy, natural, and economic combination of town and country life, and this on land owned by the municipality.

2. Crystal Palace surrounds the Central Park.

3. It is diverse architecture and design which the houses and groups of houses display is very varied.

4. Grand Avenue divides that part of the town which lies outside Central Park into two belts.

5. Because all machinery is driven by electrcity.

Unit 2 Broadacre City: A New Community Plan

1. B; 2. A; 3. A; 4. B; 5. A

1. Every citizen of the United States would be given a minimum of one acre of land per person.

2. In the three modes, the overall population density depends on the actual acreage of the surrounding parkland or greenbelt.

3. In Broadacres all is symmetrical.

4. By the most direct route from the maker to the consumer.

Unit 3   The Uses of Sidewalk: Safety

1. C; 2. B; 3. D; 4. B; 5. A; 6. A

1. Today barbarism has taken over many city streets, or people fear it has. Therefore, we need to find ways to solve this problem. First, there must be a clear boundary between public space and private space. Second, streets must be monitored to ensure the safety of pedestrians.

2. First of all, the safety of the streets is not completely maintained by the police, but by the voluntary control of the crowd, is the spontaneous behavior of the crowd. Secondly, insecurity cannot be solved by spreading people out more thinly, trading the characteristics of cities for the characteristics of suburbs.

3. The author in big American cities such as New York and Chicago, for example, looks at the basic elements of the urban structure and their function in the city life, challenges the traditional urban planning theory, to make our orientation to the development of the complexity of the urban. The city should have deepened the understanding, also for evaluating the energy provides a basic framework of the city.

Unit 4   The Four Functions of the City

1. B; 2. B; 3. B; 4. D; 5. B; 6. D; 7. A

1. From this article, I know the four functions of a city: dwelling, recreation, work and transportation, all of which will be of guiding significance for my future urban design.

2. The Athens charter divided the city into four functions: residence, work, recreation and transportation. This view still has a profound impact on the development of modern cities decades later. The large-scale development of cities in the past 30 or 40 years has led us to find that the incoordination and balance of the four functions of residence, work, recreation and transportation have led to many problems. We should also solve these problems, and this point of view will guide us to build a better future.

3. The four functions of a city must be balanced. However, excessive planning also leads to the lack of elasticity of urban development and the excessive functionalization, which further leads to the estrangement between people and the city. Therefore, planning should be flexible, leaving certain elasticity for urban development, which is also predictive.

Unit 5   Design With Nature

1. B; 2. C; 3. D; 4. B; 5. A

1. We need, not only a better view of man and nature, but a working method by which the least of us can ensure that the product of his works is not more despoliation.

2. The author leave his urban idyll for the remoter lands of lake and forest to be found in northern Canada or the other wilderness of the sea, rocks and beaches where the osprey patrols.

3. This book is a personal testament to the power and importance of sun, moon, and stars, the changing seasons, seedtime and harvest, clouds, rain and rivers, the oceans and the forests, the creatures and the herbs.

4. Our eyes do not divide us from the world, but unite us with it. Let

us then abandon the simplicity of separation and give unity its due.

5. There is an indispensable dependence between human and nature. In the future, we should consider the natural environment more when designing, rather than just focus on ourselves.

Unit 6 Accessibility

1. B; 2. B; 3. D; 4. D; 5. B

1. Accessibility, a concept used in a number of scientific fields such as transport planning, urban planning and geography, plays an important role in policy making. And this paper discusses and explains the definition of accessibility in different aspects.

2. Accessibility is defined and operationalised in several ways, and thus has taken on a variety of meanings. These include such well-known definitions as "the potential of opportunities for interaction", "the ease with which any land-use activity can be reached from a location using a particular transport system", "the freedom of individuals to decide whether or not to participate in different activities" and "the benefits provided by a transportation/land-use system".

3. Four types of components can be identified: land-use, transportation, temporal and individual.

Unit 7 Ecological Planning Methods

1. A; 2. C; 3. B; 4. A; 5. B

1. Ecological planning can solve the problem of ecological crisis. Ecological planning requires people to rationally use natural resources in

the process of urban development, maintain the renewable capacity of natural resources and maximize the protection of human living environment.

2. Regional climate, geology, terrain, water, soils, microclimate, vegetation, wildlife, existing land use and land users.

3. Microclimate means ventilation, solar radiation, albedo, and temperatures.

4. She uses the clever subtitle "Urban Nature and Human Design" to explain her approach.

5. As residents, we should protect the surrounding urban environment and use green and environment-friendly means of transportation to travel. As a planner, we should look at problems from the perspective of ecological planning, so as to create a better urban space.

Unit 8   Patrick Geddes' Visual Thinking

1. B; 2. B; 3. B; 4. D; 5. B; 6. D; 7. A

1. It shows a longitudinal section along the course of a river, from its source in the mountains to its broad mouth in the sea. The river valley is proposed as a typical basic unit for the study of the region, which encompasses natural or environmental conditions represented in the drawing through plants, with occupations or basic activities represented by tools, and with types of settlement represented by their outlines.

2. As for the image itself the following features stand out.

• It contains descriptive and symbolic elements displayed in a diagrammatic way.

• It expresses a complex concept where different aspects take part, as well as the idea of evolution in time.

• It is seemingly simple despite the complexity of its message.

3. The reality that Geddes knew is undoubtedly different from that of today, but the need to continue thinking about the nature of the city and about the mechanisms and tools for its improvement is the same. It is here where, in our opinion, lies the importance of his legacy.

Unit 9　Melbourne's Knowledge-based Urban Development

1. B; 2. B; 3. D; 4. B; 5. D

1. As a result of the spatial urban change in the city, these jobs are concentrated in Melbourne's core.

2. Economic and knowledge clusters.

3. The policies indicated a shift in the denser redevelopment of inner Melbourne which may require a substantial change in housing preferences and lifestyles.

4. Melbourne also gathers other clusters such as tourism, sports, art and culture.

Unit 10　Greenbelt of Tokyo

1. C; 2. D; 3. B; 4. D; 5. C

1. Greenbelt.

2. The greenbelt, total 136 km^2, consists of farmland and coppice woodland, was planned on a 15 km radius to restrict disordered expansion of densely inhabited urban areas.

3. The concept of the plan was to create open areas to stop the spread of fire caused by bombing and provide refuge and escape routes. The focus was to create green corridors.

4. The double-ring circular green corridors, including a greenbelt and a network of radial green corridors along trunk roads, rivers and railroads were planned to connect urban parks.

5. Only a few fluvial corridors were realized, only 4‰, 24 km², of the Ward Area is ceded as parks and open space.

Unit 11   Planning Policy The London Plan

1. C; 2. D; 3. B; 4. D; 5. B

1. It should set out an integrated economic, environmental, transport, and social framework for the development of London.

2. This process is intended to enable public participation in the plan's preparation, and reflects the principles in the EU Aarhus Convention on access to information, public participation, and access to justice in environmental matters.

3. Creating a strong policy context for growth.

4. London should take the lead in tackling climate change, reducing pollution, developing a low-carbon economy, and consuming fewer resources and using them more effectively.

Unit 12   Midtown Manhattan's Office Projections

1. A; 2. B; 3. A; 4. D; 5. C

1. New office space needed in Mid-town Manhattan to accommodate

400,000 additional office workers.

2. Priority in development should clearly go to the "soft" sites with the greatest access potential.

3. Land ownership within the presently-zoned Midtown commercial district is highly fractionalized.

Unit 13　The European Spatial Development Perspective

1. C;2. B;3. B;4. D;5. C;6. D;7. D

1. The document relates the three main policy guidelines of the ESDP:

①development of a balanced and polycentric urban system and new urban-rural relationship;

②securing parity of access to infrastructure and knowledge;

③sustainable development, prudent management and protection of nature and cultural heritage.

2. The compromise of ESPD caused many problems, which were not conducive to the development of Europe, and it should be responsible for this.

3. Given the existing regional development imbalances, the ESDP is based on the European Union's goal of achieving a balanced and sustainable development, particularly through strengthening economic and social cohesion.

Unit 14　Reinforce Hub Functions of Hong Kong

1. D;2. C;3. B;4. D;5. A

1. Because it can support the finance and business services sectors

and to enhance Hong Kong's attraction as a choice location for corporate headquarters functions.

2. These are mainly located at the fringe of the Metro Area and have good mass transport connections.

3. The Science Park has promoted development of four focused clusters in electronics, information technologies and telecommunications, precision engineering and biotechnology.

4. Hong Kong's port and airport.

5. We should continue to enhance our tourism infrastructure and make better use of our many resources to diversify our offer.

Unit 15   Residence Districts in NYC Zoning

1. C; 2. A; 3. D; 4. B; 5. B

1. R1 and R2 districts allow only detached single-family residences and certain community facilities.

2. Contextual districts generally identified with the suffix A, B, X or 1.

3. Most enclosed garage and attic space.

4. In Medium and Higher Density Contextual Districts.

5. The purpose of these amendments is to encourage development of multifamily housing.